Sigurd the
and Völund the
Smith

Tales of Two Norse Heroes

By

Katherine F. Boult

British Library Cataloguing-in-Publication Data
A catalogue record for this book is available from
the British Library

Contents

SIGURD THE VOLSUNG

OF KING VOLSUNG AND THE BRANSTOCK

In the long-past days, before the darkening of the Gods, when Odin, All-Father, came down from Asgard to mix with the men of Earth, there dwelt in Hunland a mighty king—descendant of Odin—called Rerir.

Much store had he of lands and wealth and fighting ships and brave followers, so that his name was honoured throughout the world, and all would have been well with him and his queen had they but had a son to reign after them.

Heavy of heart were they, but they prayed to the gods and offered due sacrifice, so that All-Father took council with Frigga, and, being moved by their faith, was minded to grant their prayer.

Taking an apple from Idûn's casket he bid his raven fly with it to the childless king and drop it into his hand for a sign. This the wise bird did, as Rerir sat sadly beneath an oak-tree.

" A sign, a sign! " cried the king, and springing up, he hurried to the queen with the precious fruit.

" It is the bird of All-Father," she said. " Now we know that our son will come."

And joyfully they ate the apple and set themselves to wait.

Sigurd the Volsung

But it fell out that King Rerir must go to war for the keeping of his lands and, in an evil day, he was slain and the queen left solitary. Not long did she stay behind her lord; when her son was born she lived but long enough to kiss the babe and name him Volsung, then sighed and followed whither King Rerir had led the way.

But Volsung grew great and strong beyond all other men and ruled over Hunland, as his father had done. Good fortune had he, also, in all that he took in hand and likewise in his love-marriage with Liod of the giant race. Ten sons had they, of whom Sigmund was the eldest, and one daughter, Signy, who was twin with Sigmund. And these two were the fairest and noblest of all their race.

And Signy was wiser than any woman living, yet, because of her wisdom, was ever sadness in her blue eyes, for it was given to her to know all that should befall in the days to come, and she saw before her and her people sin and sorrow and death. Because of this was her twin brother dearer to her than aught besides, since she knew that for him the future held its worst, and yet through him would come the greatest glory to her people; so that, as long as the world should last, the Volsung name would endure in honour.

Now King Volsung built for himself a house after the fashion of those times, only larger and more spacious, and the making of it was this. Near the seashore, at the edge of the great forest, stood a mighty oak-tree. Around this did the king build his feasting-hall, so that the trunk of the oak rose up in the midst and the branches came out through the roof to overshadow the house, and this tree was called Branstock. But some skalds say that it was no oak, but an apple-tree— having memory of the apple of King Rerir.

Inside, the hall was pillared with the trunks of trees, against which were stands for torches and hooks whereon each man put his weapons, so that each could seize them quickly should sudden alarm come upon them. At the upper end, facing the great south door, was the high table, and in the middle of it the seat of the king, while down either side ran other tables with benches for seats. And down the centre of the hall, between the tables, burned in winter time four fires, but in summer one only was ever alight.

Many doors opened out of the hall, some leading to the bed-places of the men, others to the rooms of the women, but the serving-men and thralls lived in other buildings around the courtyard, and only at meal-time did they come into the hall. And all around the steading was a high fence of wood in which was but one gateway, approached only by a crooked path between stakes, in order that all might be safe-warded from foes and from the beasts of the forests.

Now it chanced that in Gothland there dwelt a mighty king named Siggeir. To him came word of the beauty of Signy and of her wisdom and of her father's wealth, so that he bethought him that he would take her to wife. Fearing that King Volsung might say him nay, he made ready a great train of warships and fighting men and sailed with a large company overseas to Hunland.

And when King Volsung saw that array of warships with fearsome painted figure-heads of dragons, of eagles, and of strange monsters of the deep, and beheld the long lines of painted shields—a shield to each fighting man—hanging along their sides, his heart misgave him, for he was old and feared the wrath of Siggeir, since never was it the custom to come thus armed in friendliness. Therefore, when the king of Gothland strode up the hall of the Branstock and made his demand, Volsung answered him with fair words.

" But one daughter have I," he said, " and loth am I to wed her overseas, even to so great a king. Give me, I pray thee, time in which to think of this thing:"

Then Siggeir went forth and waited in the tent that had been set up for him, while Volsung and his sons took counsel together, whether this evil thing should be.

And all said yea, seeing that no other way there was out of this evil plight, except only Sigmund. He, knowing the future, as Signy did, saw all the woe that should come from this wedding.

" My father," he said, " nought but sorrow and grief can come of this bond. Better were it that we all should die fighting for the right than that Signy should be offered up. My life will I give most gladly in her cause."

The brethren murmured, for this was not to their mind, and, in the end, Volsung turned to Signy who, white and still, stood beside the

The Branstock

Branstock, leaning her golden head against the gnarled trunk.

"What says my daughter?" he asked. "Signy, what is thy will? Thy brethren and I fear that this mating must be."

Then the fairest Signy went to each of her brothers in turn and, looking straight into their eyes, saw there no hope of escape. But into the eyes of her father and Sigmund could she not look, since their hands were over them; and she answered:

"My father, rule me in this as in all else; yet is this Siggeir cruel and crafty, with no goodwill towards us, and I fear me that evil will fall upon the Volsungs by reason of this marriage."

Slowly and sadly, with head bent so that her golden plaits swept the floor, she passed from the great hall.

At midsummer there was a great feast made, and swift runners went throughout the land to summon the chiefs to the wedding of their king's one daughter. And King Siggeir sat on the high seat over against his host, King Volsung, and pledged him in the mead cup, passed across the central fire, as the custom of that time was—for one fire burned ever, day and night, in the hall of the Branstock. And—since it was the wont of those times to make great vows at marriage and mid-summer feasts over the Cup of Bragi—Siggeir, in his treachery, made a vow that by him the Volsungs should come to their deaths; but he spoke it not aloud, as a brave man should, but drank it silently with lowered, furtive eyes.

Sigurd the Volsung

Now, when the feasting was at its height, there strode into the hall a grey-bearded man of ruddy hue and mighty stature, who had but one eye. On his head was a hood that half covered his face, from his shoulders hung a cloak of blue-grey wool, and his feet were bare. In his hand he carried a great sword that glinted, steel-bright, in the torch-light, and none made so bold as to greet him as he strode noiselessly up the hall to the Branstock, although none guessed that this was Odin, All-Father, come to weave the fate of the Volsungs.

Amid the silence of that great company the Wanderer smote his sword deep into the trunk of the oak-tree, so that only the glittering hilt stood out. Then, turning, he said:

" There, O Volsungs, is a blade of the best. Never was a better forged. To him that can draw it forth I give it, to work the weal and woe of those that meet it. Valhalla hall is wide, welcome are the battle-slain heroes to feast therein. King Volsung, fare thee well, but not for long! "

And, without going forth from the door, that grey man vanished before the eyes of the feasters, and none knew whither he went nor did any dare to have speech of him.

Then each man, desiring to gain the sword, strove with his neighbour to be the first to touch the hilt until King Volsung said:

" Unseemly is this strife; let the noblest—our guest and son-in-law—try first, then each according to his rank and state. For it comes to me that the brand is a gift of Odin, and it will fall to him alone whom All-Father has chosen.

The Branstock

With an evil smile King Siggeir stepped from the high seat. Sure was he that he could draw forth that goodly weapon, but he strove and strove in vain and after him did also the chiefs of the Goths.

Then said King Volsung:

" Now stand forth, men of the Hunland, from the least upward to the highest. So shall I come last of all."

Again they tried, but none could stir the hilt by so much as an inch until, last but one, came Sigmund. And as he laid a quiet hand upon it, behold! the sword came forth without force.

Then a shout went up from all, for they saw that no blade like unto this had been forged since the dawn of the world. Silent and black was the wrath of Siggeir, but he smiled as he turned to Sigmund.

" Good brother," he said, " so great is thy valour that no need hast thou of such a sword. Much gold and treasure have I at home. A fair share of it shall be thine if thou wilt give me the brand, for I have a wish to keep the sword that has come at my wedding."

But Sigmund hated Siggeir too greatly to speak him fair and he answered roughly:

" What is gold to me, O king? Niddering should I be were I to part with the gift of the gods. Thy chance was better than mine to take it, since thou wert first in the trial. Why didst thou not do it? Never will I barter mine honour for gold."

Black grew the heart of Siggeir at this taunt and white was his face; but, being cunning, he smiled his cruel pale smile and hid his anger so that only Signy—being a woman—saw. But, to be revenged, he would not stay for the accustomed seven days of feasting.

At dawn next day he sought King Volsung and said:

"The wind is fair, but methinks a change will come. It were well, therefore, that we sail homewards with all speed, lest storms arise to fright my bride. Yet that there may be no discourtesy, I pray thee, O Volsung, pledge me thy word that thou and thy sons and thy chieftains will finish this feast with me in Gothland."

"That will we gladly," said the king, "to that I pledge my word for myself and my house. Within three months will we come."

But Signy, warned by her foreknowledge, went secretly to her father and said:

"Dear father, this evil mating is done and naught will I say for myself, since I must dree my weird alone. But the Volsungs live yet to glory in the strength of days and I fear me of what will befall if they enter the halls of Siggeir. Therefore, I pray thee, come not to this tryst lest much sorrow come upon us all."

King Volsung answered tenderly:

"Sweet daughter, nay. My word is given that we would go."

"Then," said she, "gather a mighty host and go in all thy war-gear."

But the king shook his head.

"As guest and not foe did I pledge my word to Siggeir, and so must it be. Better that we should suffer at his hands than that we should break a troth-word given."

Then the snow-white Signy sighed for the breaking of her heart, and without one backward glance at her home, sailed away to her life of ruth and sorrow.

Sigurd the Volsung

Now when the time drew near for the journey to Gothland, King Volsung made ready three long-ships of the best, with painted sails and fresh gilded prows, then he called his sons and jarls around him and told them all the words of Signy's warning and Siggeir's envy of the sword, and he ended:

" How say ye? Shall we sail and chance this treachery? or shall we bide at home, forsworn and niddering? "

" We go, befall what may! " they cried, and the king, well pleased, set sail with a fair wind, so that in due time they cast anchor in the haven over against King Siggeir's hall.

Seeing the sails of her father's ships black against the sunset gold, Signy the queen stole down heavy-hearted, when night fell, to the beach; for she alone knew that this would be her last meeting with her kinsmen.

" It is even as I said, my father. In the Mirk Wood hath Siggeir gathered a numberless host; as yet he knows not of your coming; will ye not sail away ere it be too late, gather an army, and come again to take vengeance on his guile? His heart is black and evil towards thee," she went on, and the last tears that she would ever shed fell fast. " Naught can save thee but to sail away this night from this dark and baleful land. Take me also, O father, take me back to our happy sunlit home."

But the old king answered quietly, as he laid his hand upon her bent head:

"My daughter, this may not be. If it be our fate to die in a strange land, then must we meet it; for never shall it be said that a Volsung turned his back on death. Wouldst thou have the maidens make a mock of thy brethren? We will finish our work as men and warriors should, that we may be sure of our welcome in Valhalla."

Then Signy wrung her hands and bade him farewell, and from that day forth did she neither weep nor smile, but, white and still, plotted ever how she might avenge her kinsfolk.

Now when morning broke, the Volsungs landed and went inland over sand-dune and bent-grass until they came to a little hill, from which they beheld the hosts of Siggeir, thick with spears as corn in a field; and, as they saw, they looked each other in the eyes and smiled until the rosy dawn made a glory of unearthly light in their faces, for they knew that this would be their last fight and their hearts were glad that soon the Valkyrjar, riding apace, would gather them in to feast with Odin in Asgard.

"Form ye the battle-wedge, my children, and break the peace strings," cried the king, and the still air was cut by the sharp snap of cord and the swish of steel on steel. Then rose the battle-cry, and down they dashed upon the waiting folk of Siggeir.

Sigurd the Volsung

Many foemen were the Volsungs minded to take with them to death. Eight times did they mow and hew their way back and forth through the thickest of the throng, until King Volsung and all his men lay dead, save only the ten sons, who were left alive and carried bound before King Siggeir, for it is not meet to kill a foe at sundown.

And the king laughed aloud and said unto his chief jarl:

"At morn they shall die and the sword of the Branstock will be mine."

"Not so, O king," the chief answered. "A fair deed would it be to send them scatheless oversea back to their own land, so that men should praise thee."

But Siggeir laughed yet more evilly and said:

"That they might come up against me once more? I would have rest and the sword of Odin. Go."

Then came snow-white Signy with unbound golden hair, praying:

"Grant me their lives but for two days, O Siggeir, since SWEET TO EYE WHILE SEEN says the rede."

Again the king laughed and said:

"Be it as thou wilt; yet it seems to me that soon wilt thou come and pray for their death."

Now he willed that his wicked witch-mother should choose the manner of their death and he stole away to her and asked her counsel, saying:

"The men have I and the sword have I also; say, mother, what shall be the fate of these Volsungs?"

She thought for a while.

"Take these ten men, my son," she said at length, "chain them to a great log in the Mirk Wood. There shall the beasts of the forest deal with them and on thee will be no blood-guilt before thy wife."

When this thing was done Siggeir went unto Signy, and she asked him for tidings of father and brethren.

"Thy father died as Volsung should," he said; "thy brethren live and wait my pleasure, as I gave thee my word."

Then to Signy came a little hope, for she said:

"Surely I may find a means of deliverance," and she called unto her a faithful servant and bade him go at dawn and bring word of the Volsungs.

But in the night came a great grey she-wolf that fell upon one brother so that he died and she ate him and vanished into the Mirk Wood.

The Faithful One, coming at twilight dawn, found but the nine and carried that word to the queen. Yet naught could she do to save them, for the eyes of watchers were ever on her, by command of Siggeir, and thus it fell out nightly until nine of the Volsungs were gone and only Sigmund, her dearest, was left.

Then she took heart and bethought her to send the Faithful One with honey, wherewith he smeared the face of Sigmund, so that the wolf, smelling at it, stopped to lick it, and in this wise was Sigmund able to catch her unawares and tear her jaws asunder so that she died. Yet in her struggle was some service, since she broke the chain wherewith Sigmund was bound, so that he was loose from the beam.

Then he knew that this wolf was a skin-changer [1] and the mother of King Siggeir, and his heart was glad, in that he had been the bane of one of this wolf race.

But of all this Siggeir recked naught, since he held the sword of the Branstock, and in time was Signy able to take counsel with her brother and the Faithful One that he should find an abiding place in a cavern deep in the Mirk Wood, where the queen might provide him with all things needful. And King Siggeir knew not that the bravest and best of the Volsung race yet lived.

Thus passed the years until the elder son of Signy numbered ten years, and she sent him unto Sigmund, that he should prove the boy to see whether he would become a hero and the avenger of the Volsungs. And Sigmund looked upon the lad and asked his business.

" My mother sends me with this word—' Try thou my son, if he be fit for thy work."

And Sigmund said:

" Mix thou the bread for our evening meal, while I fetch wood."

But when he returned, the meal-bag lay upon the ground and the child sat, frighted, in a corner.

" Why is there no bread? " he asked.

" Something moveth in the meal-bag," said the boy, " therefore I durst not put in my hand. An evil worm it is, like to a dead twig."

" Hie thee back to thy mother for a niddering," said Sigmund, " but, since thou art a king's son,

---

[1] This is the earliest reference in Norse literature to the were-wolf, which became later on such a feature of the sagas.

beware that thou speak no word of me and my dwelling."

And the boy went, betraying him not.

Then Signy sent her second son, and with him it fell out as with the elder, and Signy, white and fierce and despairing, led both into the Mirk Wood to her brother, crying:

"Slay me these children of Siggeir, since no Volsungs are they. Woe is me for the weary years that I must wait!"

Then Sigmund slew the boys, and Signy waited in the hall of Siggeir with a flaming heart and a cold white face.

Sigurd the Volsung

OF THE AVENGING OF THE VOLSUNG

THUS passed on ten slow years until the queen sent her third son, named Sinfjötli, unto her brother. He came upon Sigmund in the depths of the Mirk Wood and they two gazed at one another with silence between them. Then the ten years' old boy spoke:

"This is a wondrous thing. Here is the river and the great rock and the cavern as my mother told me, but thou art not the man who should be my foster-father, since she said that all must tremble who looked upon his face."

"And dost thou not quake, youngling?"

The boy laughed in scorn and picked up his shaft.

"Farther through the forest must I take my way. Farewell."

"Nay, thou shalt stay, for this is the token that thou art chosen, since thou hast looked on the face of the Volsung and smiled. What said thy mother?"

The lad dropped his spear.

"My mother sewed gloves upon my hands through skin and flesh, yet I made no sign; then did she peel them off so that the skin came with them, yet I cried not out. Then said she: 'Go tell Sigmund that I send a man to work his will,' and here am I."

The Avengers

Then Sigmund tried him with the meal-bag as he had done with the others, and when he came at evening the bread was ready.

"How comes this?" he said; "didst thou find aught in the meal-bag?"

"Ay, truly," answered Sinfjötli, "something alive there was in the meal, but I kneaded it quickly and baked it, and now the thing is alive no more." And he laughed.

The heart of Sigmund laughed also, for he knew that the avenger had come.

"No bread wilt thou have to-night," he said, "for a deadly worm hast thou baked therein."

But he might eat of the bread since venom within could not hurt him, while Sinfjötli could but bear venom without.

Then did they fare far and wide through the forest, taking vengeance on King Siggeir's men when they could, but keeping far from his hall, since Sigmund judged the lad yet too young for the plan he had in mind. Together they did mighty deeds, for Sinfjötli was wild and savage beyond other men, and Sigmund, looking on him, feared that, with the Volsung strength, he had gotten the evil heart of Siggeir; yet never were his deeds crooked, and he ever spoke the truth, since fear he had none.

Now it fell out that, seeking plunder, they one day came upon a hut where lay two men asleep. Upon their arms were great gold rings, whereby Sigmund knew them to be the sons of kings, and near their heads hung wolf-skins, for they were spell-bound skin-changers — like the mother of

Siggeir—but only every tenth day might they come forth as men.

Then Sigmund and Sinfjötli put on them the wolf skins, and, with them, the wolf nature, and no more could they come out of them, and Sinfjötli said:

" It seemeth to me that with the wolf skin, I take the way of wolves, for I know the voices of the Mirk Wood and much that was hidden heretofore."

" So it is," said Sigmund; " to me is thy speech plain, but to earthmen would it be but a wolf's howl. Now see. Each of us will take his way alone through the forest and seek out men. But because thou art still young and not come to thy full strength, this must thou do. If there be but seven men together, fight them; but if there be more than seven, then shalt thou call on me for help in this wise and I will come."

And Sigmund lifted up his head and howled, that the youth might know the cry, and thus they parted.

Then went Sinfjötli, and meeting eleven men, he slew them all after much fighting; but he was sorely wearied therewith and lay hidden to take rest. To him, softly padding through the wood, came Sigmund, and as he came he passed the eleven dead men.

" Why hast thou left me uncalled? " he asked, looking down on the panting wolf. And Sinfjötli laughed.

" What are eleven men? " quoth he.

Then came the wicked wolf nature upon Sig-

mund and he sprang at the youth and bit him in
the throat so that he lay sore hurt—even unto
death—then, seeing the wound, his manhood came
again and he sorrowed grievously in that he had
slain the boy, and he cursed those wolf-skins and
their evil nature. Yet in no way could he mend it,
so, with sore labour, he bore him back to the cave
and sat by him to watch.

And as he sat, there came two weasels fighting,
so that one slew the other; then the whole one,
running into the thicket, returned bearing the leaf
of a herb that it laid upon the dead weasel so that
straightway it sprang up whole and sound. Then
did Sigmund follow to see if, perchance, he also
might find that wondrous herb, but, as he searched,
there came to him a raven bearing a leaf in its
beak. This he took and drew it across Sinfjötli's
wound, back and forth; and as he did so, breath
came back to the lips of the boy and light to his
eyes, so that he arose in full life.

Then did Sigmund know that the raven was the
bird of Odin and that Sinfjötli was chosen for the
work he had in mind. With a glad heart he bore
the ten days of their wolf-hood, and when they were
ended, he and the youth made a great fire and
burned those skins so that never again should any
man don them to his hurt.

Then, on a day chosen of Sigmund, these two
went forth from their earth-house and the Mirk
Wood to the hall of Siggeir the king, and they hid
themselves among the barrels of ale that stood
in the porch, while the Faithful One went and told
Signy that they had come.

Now two little children, the youngest born of Signy and Siggeir, played in the hall with a golden ball, and the ball rolled to the place where the two were hid. Thither came the children seeking it, and beheld the warriors sitting grim and still, and they ran to the king, as he sat on the high-seat, and told him, saying: " Behold the great kings who sit silent at the back of the porch, O father, in their golden helmets and shirts of glittering mail. Have they come at last, as our mother's stories said they would some day? "

And the king turned, glowering on Signy, but the queen's blue eyes flashed fire, and her head rose proudly, for she knew that her vengeance was nigh.

She looked not on her husband, but, taking a child in either hand, she stepped from the high-seat, and swept down the hall, while all men's eyes went after her, wondering. And, standing by the porch, she cried aloud:

" Come, brother, slay me these betrayers."

" Nay, sister," answered Sigmund, " though thy children betray me, I slay them no more."

But Sinfjötli picked up the children and cast them down dead.

Then Siggeir stood up and called to his men: " Slay me these men in the porch."

And the fighting men ran together in haste, and after much toil took Sigmund and Sinfjötli— for they were but two against many—and bound them. And because it was not fitting to kill captives at sundown, the cruel king bethought him what would be the hardest death for them to die, and he bade the thralls dig a deep hole in the ground, and therein he set a big flat stone on its edge, so that the hole was divided into two parts, and he set Sigmund on one side of the stone and Sinfjötli on the other, that they might each hear the other's voice, and yet be parted.

And while the thralls were turfing over the hole, Signy came quickly and cast down a large bundle of straw to Sinfjötli. Then was the turfing finished, and the two buried in black night within the barrow. Then Sinfjötli cried:

"The queen has sent us meat here in the straw, and thrust in the meat is thy sword, for I know the touch of the hilt."

"Then let us saw the stone," said Sigmund, "for naught blunts my good sword."

So Sinfjötli drove the point of the sword hard through the stone, and Sigmund caught it and they sawed, as stonecutters do, until the stone fell asunder.

Then was it easy for them to cut a way out through wood and turf, and they piled faggots around the hall of Siggeir and set fire to it; and Siggeir from within cried:

"Who hath kindled this fire?"

"I, Sigmund the Volsung, that thou mayst know that a Volsung yet liveth."

Then he called above the roar of the flames:

"Signy! Signy! beloved sister, come forth. Thou hast dreed thy weird. Come forth and receive atonement for the sorrows of thy life in Gothland."

So Signy came forth; her blue eyes blazed in

the fierce firelight, but her golden hair was white with the sorrow of all the years.

She kissed her brother and her son, saying:

" Dear brother, through thee hath my vengeance come, and I care to live no more. Sadly did I wed King Siggeir, but gladly will I die with him. So fare ye well."

And with head held high, and no backward look, she swept straight into the flames, and so died with King Siggeir and his men.

Sigmund's Death

THE DEATH OF SINFJÖTLI AND OF SIGMUND

Now was there naught more to do in Gothland, and Sigmund gathered men and ships and sailed home to Hunland to live in the hall of the Branstock.

He was a great and wise king, but never did the sorrow of the Volsungs' ending die out of his heart. He took to wife a jarl's daughter called Borghild, who hated Sinfjötli, both for the love Sigmund bore him, and also because, in fair fight, he had slain her brother. She begged the king to send him away out of the land; but he would not, and gave her instead great stores of gold and amber and jewels as were-gild,[1] for in those days the price of a life was paid to the nearest of kin.

Now, in honour, Borghild, having taken the price of blood, should have been silent and have pardoned the slayer; but she thought only of vengeance, and made a mighty funeral feast, bidding thereunto all the great ones of the land.

When they were seated she carried horns of ale and mead to those whom she wished to honour, and amongst them, one to Sinfjötli, saying:

" Drink, fair kinsman."

Sinfjötli, looking into her eyes, beheld their guile, so he said:

[1] Blood-money.

"Nay, it is a witch-draught, and I drink not."

"Give it to me," said Sigmund, and he drained the horn, for no venom could harm him.

Then came the queen again to Sinfjötli, saying: "Come, drink! Must other men drink for thee?"

He took the horn, and looking into it, he answered: "It is a baleful drink."

And again did Sigmund take the horn and drain it, while the angry queen tossed her head. Yet a third time she came, saying:

"If thou hast the heart of a Volsung, and art no niddering, drink!"

Now, no true man can be called a niddering, even by an angry woman, so Sinfjötli took the horn and drained it, saying:

"Venom is in the drink," and as he spoke he fell dead, and so great was the shock of his fall that the Branstock swayed.

Sorrowing unto death, Sigmund rose up with his sister's son in his arms, and strode away through the forest until he reached a lonely fjord, where sat a man in a little boat, and the man had but one eye.

"Those who sent me told me that a king and the son of a king would come," said the man, "but, since the boat will hold but two, walk thou round by the shore."

And behold, as Sigmund turned, boat and steersman vanished away; so the aged king knew that this was Odin come to take a Volsung home.

The evil Borghild did he drive away and, after

wandering awhile, she died and left him free to wed again.

There lived, near by, a king who had a fair and wise daughter named Hjordis and a son Gripir, who had foreknowledge. She was sweet and full of love, and it seemed good to the aged Sigmund that he should pass the evening of his days in peace with this gentle maiden.

He, therefore, journeyed into her father's land, and so did also a younger king, Lynge, son of Hunding, who willed to wed her.

Fearing strife, her father said to her: " Hjordis, my word have I given that thou alone shouldst choose thy husband. Say, therefore, wise daughter of mine, which wilt thou take? "

" This is a hard thing," she replied. " Yet will I choose King Sigmund. Though he be old, still is he the greatest of heroes, and of the Volsung race."

So were Sigmund and Hjordis wed; but the young king, Lynge, went away with rage in his heart, and he and his brethren gathered a great army for the undoing of King Sigmund.

Then sailed they to Hunland, and Sigmund gathered his men by the sounding of King Volsung's horn, which in peace time hung on the Branstock.

And now began the most awesome fight that had been since the death of King Volsung; but ere it began, Hjordis, with her tire-maiden and much treasure, was hidden in the forest.

All day the fight went forward, and old though Sigmund was, none could prevail against him.

Naught could one see but the swift flash of the sword that no man might break, as he hewed his way through the throng, his arms red with blood.

Now, when the sun was at its setting, there came up against Sigmund a stranger in a blue-grey cloak; one-eyed and grey-bearded was he, and he carried a spear in his hand. And as the king's sword smote against the steel, behold the good sword split in two pieces and its fortune was gone.

Then the tide of battle turned, and the Volsung's men fell fast until all—even the father of Hjordis—were dead or sore wounded. But Sigmund, still living though stricken with death, lay upon the field. And through the darkness crept the gentle, hapless queen seeking her lord. Kneeling beside him and wiping the death-dews from his face, she asked: "Canst thou not be healed even now, my king?"

"Dear wife," he answered, "nay, and I would have it even as it is. I have lived long, and with my sword has my fortune left me. Nor does Odin will that I should live, for he himself it was who broke my sword; and to him shall I journey, riding straight to Valhalla gates. So to another must I leave it to avenge thy father—to a mightier than I. Thou shalt have a son; care for him well and save for him the two pieces of my sword; thereof shall a noble weapon be made that shall be called Gram, and Sigurd shall wield it. He shall be the last and noblest of our race, and while this earth lives shall the name of the *Golden Sigurd* be known. Now fare thee well, dear heart, for I weary with my wounds and fain would feast with Odin."

Then Hjordis kissed him and laid his head upon her knees, and so sat, with her handmaiden beside her, until the daybreak; and, as the first light came, the great king looked up into her eyes and smiled and died.

But King Lynge sought through all the land for Hjordis and her treasure, and finding them not, he took the land and harried it.

Sigurd the Volsung

THE LAND OF HJAALPREK THE HELPER, AND THE BIRTH OF SIGURD

Now the day was come, and Hjordis the queen arose wearily and, looking over the sea, saw many ships. Fearing another foe, she said unto her maiden: "I dread more strife and am here helpless with none to defend me; change thou, therefore, thy raiment with me that none may know me for a king's daughter." This they did as the seamen came up from the ships; at their head was Alv, son of Hjaalprek, King of Denmark, who inquired of the women what this slaughter meant. Then Hjordis answered and told of the great fight, and the Viking prince marvelled at her sweet low voice and clear words, which were not those of a bond-woman, and spoke her fair so that she believed in him and showed him where the Volsung treasure lay hidden from the foemen who sought it.

When all was gathered together and put upon the ships, and a barrow raised over the dead kings, the prince asked the bond-woman:

"Wilt thou, O Queen, with thy handmaiden wend back with me to Denmark?"

And the false queen looked to Hjordis to answer, and she said:

"If peace dwell in thy land, O prince, thither

will we gladly wend, for we are but weak women and are weary of strife."

Then they sailed away, taking Gripir also, and the prince took the helm of his ship and talked with the women as he steered; and every hour he wondered more that the maiden should be so much wiser than her lady.

After a fair voyage they came to the low shores of Denmark, where fir woods come down to the sands by the sea, and Alv led the maidens before his father, Hjaalprek, and his mother, who bade them welcome, and treated them with honour.

Then Alv spoke to his parents of his thought that Hjordis was no thrall's daughter, and the wise old king made a plot to catch them and to learn the truth.

As they sat together at eventide around the fire in the hall, Hjaalprek asked the maiden: " How knowest thou when dawn is nigh? "

" Whereas," she answered, " I milked the kine when I was young, now wake I ever at the self-same hour before the dawn."

Then Hjaalprek the king laughed a mighty laugh, being well pleased.

" Kings' daughters milk not kine," he cried, and turned him to Hjordis, and put to her the same question.

" I know," she answered heedlessly, " by the little gold ring given me by my father, which groweth ever cold at the dawn of day."

Then spoke Alv:

" No bond-maid art thou, but a princess; why hast thou dealt doubly with me? Hadst thou spoken truly thou shouldst have been as my sister."

Hjordis felt shame for her deceit, and she knelt before the kind old king and queen and said:

" The wrong is mine, but pardon me, I pray. Bethink you! I, the widow of Sigmund the Volsung, was alone with this my maiden in the midst of my enemies, even my young brother lost for the time. I knew not but that your son was their helper; how could I tell how good a friend he would be to me? "

And the widowed queen turned and smiled upon Prince Alv, so that he loved her the more; and he stood forth in the midst of the hall, leading her by the hand, saying:

" For thy beauty and wisdom do I love thee, Hjordis, and when thy days of mourning for the great king are past, then shalt thou be my wife."

Before many weeks were over, the son of Sigmund and Hjordis was born; they carried the child upon a shield to King Hjaalprek, who rejoiced greatly over him and, calling for water, named him Sigurd, according to his father's will. Also, seeing the child's keen bright eyes, shining like stars, he foretold that throughout all the earth no man should be his equal.

Thus in the midst of peace, love, and honour, grew Sigurd. Brave and true-hearted, he scorned a lie, nor ever sought his own advantage. Yet withal, he was so gentle that little children ever ran to him and loved him. Yet could he fight, and was he ever foremost in warlike sports, bearing in mind that he must be the avenger of his father.

The Birth of Sigurd

The wise old king chose for him a teacher to show him all those things that princes should know; so was he learned in all games of skill, in speech of many tongues, in metal work, in woodcraft and in shipcraft.

This teacher was Regin the master-smith, son of Hreidmar. A strange being was he, misshapen yet not a dwarf, silent and glowering unto all save only Sigurd; skilled in runes, in the lore of many lands and in metal work, so that the people whispered of his kinship to the underground folk, who have all metals in their keeping. But Hjaalprek knew not that he was full of guile, and that throughout the years of Sigurd's growth he plotted how he might use the lad for his own wicked ends, and be his undoing. Thus one day he said:

" It is shameful that thou hast no horse. These kings treat thee as their foot-boy. These kings, forsooth! in whose land is peace, and who go not out to fight."

" That is false," said Sigurd hotly, " and thou knowest it. If I need a horse I have but to ask. The kings are beloved of all and need not to fight. Yet if fighting were toward, father Alv would do his part."

And he went angrily out of the smithy. But after some months 'he went to King Alv and begged a horse of him, and the king said:

" Go choose thee one from the herd by Busilwater; they are the best, and all that is mine is thine, brave son."

Sigurd blithely thanked the king, and took his way to the meadow far up the woods, where the

Sigurd the Volsung

Busil-water ran. On the way he met an aged man, with a long grey beard and one eye, who asked whither he fared.

" To choose me a horse, O Ancient One. · If thou art a judge, come with me to help my choice."

And the old man journeyed with him, telling him of his father Sigmund, and his forefather Volsung, whom the Aged One had known. Then Sigurd knew that this must be one of the god-folk, to have lived so long.

As they talked, they came to the green meadow where the horses were, and the old man said :

" Now, will we drive the horses through the river of roaring water, and watch what will betide."

And the force of the water, rushing down from the mountains, frighted the horses, so that they turned and swam to land again, save one grey horse with a broad strong chest, who feared naught. He alone swam to the far side, and there landed, neighing and stamping in pride, then plunged into the torrent once more and swam back to the Ancient One and Sigurd.

" This one must I choose: is it not so? " asked the lad; and the old man answered : " Thou chooseth well, for he is of the race of Sleipnir, All-Father's horse, that never tires," and, as he spoke, he vanished away; and Sigurd knew that this must be Odin himself.

Then he took the horse, which he named Grane, and went back to the hall of the kings well pleased, and they and Hjordis rejoiced with him.

But after a time, crafty Regin went yet further

with his plan, and he asked: "Where is the treasure of thy father, the Volsung?"

"It is in the treasure-room of Queen Hjordis," Sigurd replied; "it is a fair treasure, but I have heard of greater, gathered by some kings."

"Why is it not thine?" asked Regin.

Sigurd laughed and said: "What should I, a boy, do with this treasure? It is naught to me, and there is no magic in it, else might I desire it."

"And wouldst thou have a magic treasure?" asked Regin keenly.

"I know not," the lad answered carelessly. "A great hero can I be without aid of magic. It was idle talk."

"But if I could help thee to great treasure and glory, wouldst thou refuse?"

"Why surely, nay," quoth Sigurd; "is it not for glory that the Volsungs live?"

"But a little way hence on the waste of Gnita Heath it lies, a treasure greater than has been seen in the world, and over it doth Fafnir keep watch and ward."

"Of this worm, Fafnir, have I heard," said Sigurd; "more evil and mighty is he than all other dragons."

"Nay," quoth Regin, "an over-great tale of it do men make; he is but like to other worms, and a small matter would thy forefathers have made of it. But little of the Volsung spirit has fallen to thy share!"

"Spirit have I," said Sigurd hotly, "but I am not yet come out of childish years. What is in thy mind that thou shouldst flout me thus unjustly?"

Sigurd the Volsung

Regin answered not for a while, then he said:

" Come, then, and I will unfold to thee a tale that hitherto no man has known."

And the aged man and the stripling laid them down under a spreading oak in the greenwood, while Regin told this wondrous story.

The Rhine-Gold

THE RHINE-GOLD

HREIDMAR, king of the dwarf-folk, was my father, and brothers had I two. Fafnir, the elder, was having and grim; ever would he take the best, and of the best all that he could, for he loved gold. Otter was the second, and his will was to be ever fishing, so that Hreidmar gave him the gift of changing into an otter, and thus he spent most of his life on the river rocks, landing only to bring fish to my father. I was the third son, a weak, misshapen thing, but with, as thou hast seen, the gift of runes, and cunning in all metal work.

It fell one day as Otter slumbered beside a half-eaten salmon, that Odin and Loki passed by. Now Loki, the wicked one, would ever be at evil, and he caught up a sharp stone and hit Otter, so that he died. Rejoicing, he flayed off Otter's furry skin, and, casting it over his shoulder, went on with Odin to Hreidmar's hall—a golden house of beauty that I had built for him. Hreidmar, knowing the skin for that of Otter, seized the gods and cried:

" By the beard of Odin, ye go not forth until ye pay me, in were-gild for my son, as much gold as will cover his skin inside and out."

" We have no gold," said Loki.

" The worse for thee," said Hreidmar, for he

was grim and hard, and angered that no more would Otter fish for him.

Loki the crafty thought awhile; then he said:

" If thou wilt give me leave I will go take Andvari's gold."

Now Andvari was a dwarf, who lived in Otter's river, under a waterfall that was called Andvari's Foss. He guarded a great treasure that he had stolen long years before from the Rhine maidens in the Southern land, but that history belongs not here. For the most part he took the shape of a pike, so that, with the greater comfort, he might guard his treasure.

Hreidmar gave leave, and Loki hurried down to Ran the sea-goddess, and begged her magic net. This she gave, and Loki, casting it under the foss, drew forth Andvari the pike.

" What ransom wilt thou, evil one? " cried Andvari in terror.

" All thy ill-gotten gold, O dwarf."

" That shalt thou never have."

So Loki hung the net of the goddess upon a tree, and sat down to watch the great pike struggling and gasping. At last Andvari said feebly:

" Put me back in the foss; thou shalt have my gold." And he brought it forth.

But Loki, as he gathered it up, espied a gold ring around his fin, and said:

" Thy red-gold ring must I have also."

Then Andvari shrieked with rage, and threw the ring at him, cursing him and the Rhine gold and all that should own it.

" To every man that owns it," said he, " shall it

be a bane and a woe, until it return to the Rhine daughters. To each holder of the ring shall come an evil death, and because of it the hearts of queens shall break, and the Twilight of the Gods shall come."

And he plunged into the foss and was seen no more.

Back went Loki to the House Beautiful and cast the gold at my father's feet; but the bane-ring gave he to Odin. Now this ring was that one that Odin had laid on the pyre of Baldur dead, and to it was given the gift of making, every ninth night, eight rings equal in weight to itself.

Then was the fur spread out and covered with gold, first on the one side, then on the other, till but one hair was uncovered. And Hreidmar spake:

" There is yet one hair showing."

The gods looked one upon another; then Odin drew the ring from his arm and cast it upon the skin, so that the hair was hidden. Then Loki mocked and sang:

> " A great were-gild hast thou!
> But thou and thy son
> The bane shall it be of ye both."

And the gods departed.

Then Fafnir, looking covetously on the gold, slew our father for it, and me, being weak, he drove away; and, taking it to a secret place on Gnita Heath in the Desolate Land, he changed himself into an awful dragon, the better to guard it; and there is no worm like unto him, for he is made up of sin and evil. So I have no part in that which is rightfully mine, and I would that thou shouldst win it for thyself, O Sigurd.

" But wherefore," asked Sigurd, " shouldst thou not fight and win it? "

" What chance hath a weakling against that great worm? " said Regin. " Besides, my doom is that I should be slain by a beardless youth."

Then up sprang Sigurd and cried:

" Forge thou me a sword of power, and when my father is avenged, even then will I go up with thee against thy brother and get thee the gold thou cravest."

And Regin rejoiced that his plan worked, and they went back to the hall of the kings, speaking of the sword that should be forged; but Regin told not Sigurd of the helmet of darkness and the mail coat of gold that were with the treasure of Andvari.

So after some days he put a sword into the hands of Sigurd, and the lad, looking at it, laughed in mirth.

" Why dost thou laugh? " asked the master.

" Because thy hand hath lost its skill. See! " and Sigurd smote the sword upon the anvil so that it flew in pieces.

Then Regin forged yet another, and said:

" Hard art thou to please. Mayhap this may be to thy mind."

And Sigurd looked at it, and smote it upon the anvil so that it split in half. Then he looked keenly upon Regin and frowned, saying:

" Mayhap thou also art a traitor like thy kin.

The Rhine-Gold

Is it thy will that Fafnir should slay me, that thou forgest me swords of wood? Canst thou do no better than that? "

And he turned from the smithy and went to his mother; but Regin was angered at his words and hated him.

Queen Hjordis sat in the women's room broidering with her maidens, when her son cast himself down by her side and, seeing that he spoke not, she said:

" What ails my son? Needs he aught that the kings and I can give him? "

".All love and much honour have I ever from thee, mother mine, and for this I owe thee all thanks and obedience. Yet one thing I lack. Have I heard aright that thou hast the sherds of the sword that my father, Sigmund, gave thee at his death? "

" It is true," Hjordis said, but her heart was sad, for she knew that their parting time had come.

" Fain would I have them, for with no sword but Gram can I do my life's work."

Then she led him to her treasure-chamber, and from its silken coverings in the old oak chest she drew the pieces of the sword, glittering and bright as in the day that the Wanderer smote it into the Branstock, and she gave them to Sigurd with a kiss.

Blithely went the lad forth, but Hjordis looked after him, wistful, yet rejoicing in that the prophecies of Sigmund and Hjaalprek were to be fulfilled, and that her son, with the eyes like stars, should be the hero of all the ages.

Sigurd the Volsung

At the smithy door Regin met him, frowning.

"Will naught serve thee but Gram?" he asked in wrath.

"Naught but Gram!" Sigurd said, and laughed. "Gram shall slay the Serpent; take it and do thy best."

Regin took it and shut himself for many days into the smithy with his men and, after much labour, the sword was wrought; but the smiths told how, as Regin bore it forth from the forge, fire ran adown its edge. Regin looked at it and said:

"Well know I that I shall die by the sword of a youth, but, if it be by Gram, then am I content: for I am weary of the length of days that have dragged on since I forged this blade for Odin the Wanderer."

To Sigurd, waiting at the smithy door, he gave the sword, saying sullenly: "If this be not good, then is indeed my craft gone."

Then ran Sigurd joyfully down to the stream and cast therein a lock of wool and, as it floated down, it met the edge of Gram and the lock became two, and Sigurd laughed again.

Then said Regin: "Bethink thee, now thou hast a sword to thy mind, of thy promise to go up against Fafnir?"

"That will I gladly when I have avenged my father on the Hundings," said the lad.

Then the kings made ready many ships, and Sigurd was chief over them, and they sailed to the land of the Volsungs, and in a great battle slew King Lynge and the Hundings, and added that kingdom to the lands of Hjaalprek the Helper.

The Rhine-Gold

And ever in the thickest of the press gleamed Gram.

Now, when he was come home some time, Sigurd grew weary of quiet, and Gram rattled in its sheath under the peace-strings, as it hung on the wall over Sigurd's seat.

So he went to Regin, who sat wearily by the smithy fire; he turned not as Sigurd entered and, drawing up a stool, sat by him. After a while the lad spoke:

"To-morrow will I ride with thee to the Waste, Regin, if thou wilt; maybe I shall slay thy brother."

"Two shall go forth," said Regin gloomily, "but neither shall return."

"No matter," quoth Sigurd, "we will try our best for the hoard."

And that night he went unto his Uncle Gripir and learned from him all that should befall him in the future; though Gripir was sore troubled and scarce would speak at the outset, yet in the end he told unto Sigurd all that his life should bring.

Sigurd the Volsung

THE SLAYING OF THE WORM

ERE the dawn Sigurd arose and, going silently, he went to his mother and kissed her gently, for he knew from Gripir that he should see her no more; then, saddling Grane, he rode forth to the Lonesome Waste, with Regin at his side.

Ever inland and upward they rode as the days went by, leaving meadows, trees, and all green things behind. At last they came out upon the Waste beside a mountain torrent, where Fafnir was wont to drink, and Sigurd traced the broad band of slime that he made as he crawled back and forth.

" Surely," said he, " this dragon brother of thine is greater than all other ling-worms, from the breadth of his track? "

" Nay, not so," said Regin. " Dig thou but a pit in his path and sit therein, then canst thou stab him from beneath. As for me, since in naught can I help thee, I will get me to a place of safety," and he rode down the rocks.

Then Sigurd put Grane in shelter, and turned to cross Gnita Heath; and, as he went, there met him a grey-beard with one eye, who asked him whither he went and what to do, and Sigurd told him.

" That counsel is evil," said the Ancient One; " bide thou here and dig many pits, else into one

will the dragon's blood flow and drown thee as
thou standest."

And ere the youth could answer he was
gone.

So Sigurd spent the night in digging pits in the
path of Fafnir; and at early dawn, as he sat in the
largest, he felt the trembling of the earth, and knew
that Fafnir was nigh.

Snorting and spitting venom as he went, the
great Worm crept slowly on, fearing naught and,
as he passed over the pit, Sigurd thrust up Gram
with all his strength behind the dragon's left
shoulder, and drew it forth black to the hilt; and
Fafnir's blood gushed forth and covered Sigurd
as he stood, save only in one spot between his
shoulders where a dead leaf had lighted. Then
he leaped from the pit and stood afar off, as the
mighty Worm lashed out in the pain of his death-
wound, crying, "Who art thou, and whence? thou
that art the undoing of Fafnir."

But Sigurd, mindful that Fafnir might curse him
if he told his name, answered: "Nameless am I,
and born of nameless folk."

"Ah," cried Fafnir, "shame that I should be
slain by a liar. He should be a hero that bringeth
me doom, yet can a hero lie?"

Then was Sigurd shamed, for he ever told the
truth, and he said:

"I am Sigurd, son of Sigmund the Volsung,
and no liar. Tell me of the days that are to come
to me."

For all men believed that to the dying was the
future clear, and Sigurd willed to see if the words

of Gripir and Fafnir were the same. And Fafnir
spoke:

" I see bane unto thee from the gold, Andvari's
hoard, and from the fatal ring. Take thy horse
and ride away and flee from the evil. Yet shall
we meet and fight again in the day of the Destruc-
tion of the Gods, thou Golden Sigurd."

" Nay," quoth Sigurd, " for thy gold I came,
and without it will I not go. Without gold cannot
man live."

Then Fafnir poured forth words of ruth and
wisdom; and as the sun went down he quivered and
lay a chill grey heap upon the Waste, and the
sunset light shone upon the bright hair of the
Golden Sigurd as, sword in hand, he looked down
on the fell mass.

Then came Regin, who had watched from afar,
hastening to greet Sigurd.

" Hail, lord and conqueror! " he cried; " hence-
forth shalt thou be known throughout the ages as
Fafnir's Bane."

" Small aid wert thou," laughed Sigurd, " hiding
while I fought."

" Yet," said Regin grimly, " were it not for the
sword I forged thou hadst now lain low before
Fafnir. And, since he was my brother and thou
hast slain him, for atonement shalt thou roast me
his heart with fire, that I may eat it."

" That will I," said Sigurd, and he set to gather
sticks in the gloaming while Regin slept and the
birds gathered round, and he set Fafnir's heart
upon a stick to roast.

When it should have been ready, Sigurd laid his

fingers upon it and the fat, hissing out, burnt them so that he put them in his mouth to cool; and behold, straightway he knew the words of the wood-peckers that chattered as they hopped around.

The first said:

"Thou foolish Sigurd to roast for Regin. Eat thou the heart and so become wisest of men."

The second said:

"Thou guileful Regin, that wouldst betray the trusting youth."

The third said:

"Smite thou the guileful one, Sigurd, and become thyself lord of the gold."

The fourth said:

"That is good counsel, to take the treasure and hie over Hindfell to sleeping Brynhild."

The fifth fluttered and said:

"Sigurd is a fool if he spare him whose brother he has just slain."

Then up sprang Sigurd, saying:

"Regin shall be no bane of me. He shall follow his brother."

And he smote Regin with Gram, so that his head rolled away.

Then the birds rejoiced and sang glad songs of Sigurd's journeyings, and of Brynhild over Hind-fell, whom he should find, while Sigurd ate part of Fafnir's heart and saved the rest.

Then he leapt on Grane and rode by the dragon's slimy trail until he came to the great cavern; and, although it was now night, the cavern shone with a light as of day, by reason of the golden shine of the Hoard.

Sigurd the Volsung

So he set Andvari's ring on his arm and dight upon his body the golden mail and upon his head the helmet of darkness, and, putting the Hoard into two chests, he fastened them upon the back of Grane, being minded to walk himself because of their weight. But Grane stirred not, and Sigurd was troubled what he should do, for even he dared not smite the horse. Then he looked into the eyes of Grane and knew what was in his mind, so he gathered up the reins and leaped upon his back, and the grey horse tossed his mane for joy and galloped over the Waste, turning southward, steady and untiring.

Brynhild

By stony ways rode Sigurd southward towards
the Frankish land and, as he came over Hindfell, he saw before him a mountain whereon a great
fire burned, and in the midst of the fire a castle
with a floating banner, with shields around the
towers.

And he climbed that mountain until he came
close to the fire, and the crackling heat of it fanned
his curls. Then he cried unto Grane, and the
brave grey horse, with one mighty spring, leaped
through the flame and stood at the castle gate and
Sigurd, looking back, saw only a line of grey ashes
where the fire had been.

The castle door stood wide and Sigurd, with
Gram unsheathed, strode through the empty
courts. Upon a rock in the inmost hall lay a man
in full armour, his face covered by his visor. Then
Sigurd cried aloud:

"Arise! I am Sigurd."

But the figure moved not; so, with the point of
Gram, he loosed the mail coat and flung it off and
the string of the helmet and cast it aside, and behold! there lay before him, in deep sleep, the fairest
woman he had ever seen. Gold was her hair as
the hoard of Andvari, white was her skin as the

froth of sea waves, and her opening eyes were blue as a mountain tarn.

"Who waketh me?" she asked, low and soft as in a dream. "Me, in whom Odin, All-Father, set the sleep-thorn because I did as he willed not. Is it thou, Sigurd, son of Sigmund, with Fafnir's Bane in thy hand, and Fafnir's helmet on thy head?"

"It is I," he answered; "tell me thy name."

"I am Brynhild, Valkyrja of Odin. Against his word did I give the victory to the man he would not; therefore did he strike me with his sleep-thorn and lay me within the fire-ring. And this doom is laid upon me, that never more shall I choose the slain; that now am I mortal and must suffer my tale of woe, even as the children of men; that I shall wed but a mortal and bear the bitter things of life. But this have I vowed—since I must wed—I will lay my hand only in that of a man who knows no fear."

"Surely," said Sigurd, "thou art both fair and wise. Tell me of wisdom and love during this day that I may spend with thee."

And Brynhild told him of the secret runes of the gods and of many things hidden from men. Through this and through his knowledge of bird-speech became Sigurd wise above all men.

Now when the day was ended the Volsung stood before the Valkyrja, and in that deep voice like unto the music of a mountain torrent said:

"I am he that knoweth no fear. I swear that thou, Brynhild, art near to my heart, and none will I wed but thee."

Brynhild

And by the two hands he held her, looking deep into her eyes, as she answered:

"Thee do I choose before all the sons of men, O Sigurd."

So they plighted their troth and drank of the love drink, and he set upon her arm the red-gold ring of Andvari. And thus began the Valkyrja's sorrow; yet, having the love of the best of the Volsungs, would she not change it for aught of mortal joy.

Now when the new day was come, Sigurd arose and clad him in the golden armour of the Hoard, whereon was drawn the image of that dragon which he slew, and upon his red-gold hair he set the helmet with its dragon crest.

"Fair love!" he said, kissing Brynhild between the eyes, "I must fare forth to do the deeds that await me and to meet the fate that is set. Yet ere long will I seek thee in thy sister's home at Hlymdal, and at my coming shall we have much joy."

But Brynhild sorrowed and answered low:

"Woe is me, my hero; for thee and me will be no bridal until our death-day join us. Thou wilt wed a daughter of the southland folk. We must dree our weird apart."

Then Sigurd laughed and kissed her, saying:

"Sweetheart, thou art sad at our parting. Thou, daughter of the gods, knowest full well that what will be must be, and naught can mortals change when the Nornir have spoken."

And again he kissed her and rode down through the valley—Golden Sigurd in the sunbeam glint— and Brynhild watched him till she could see him no more. Then she turned and wept the tears of a mortal woman for the first time, and made her ready to go to Hlymdal.

Now in Hlymdal dwelt Heimar, a noble chief, who had wedded Beckhild, sister to Brynhild and Atli. Thither to his home came Brynhild to pass the time of her waiting for Sigurd to come.

One day as she sat in her tower there came, running, her maiden, who said:

"See! who cometh over the hills with this train of men and horses?"

And Brynhild looked forth and sighed heavily. "It is Gudrun, daughter of Gjuki, King of the Niblungs, and her coming brings me woe."

Then went she down to greet Gudrun, fairest maiden of Frankenland, and give her welcome. But Gudrun was sad and heavy of heart as they sat in the high - seat together, and Brynhild said:

"Canst thou not laugh and be merry as we used of old, O Gudrun?"

"That can I not for dreams that trouble me," answered Gudrun; "even for that am I come to thee, that thou mightest unravel all for me."

And Brynhild led her to her tower and set her in the high-seat, saying unto her, "Say on;" but she sat herself at Gudrun's feet with hidden face.

And Gudrun spoke:

"Thou and I, Brynhild, were with other women at the hunting of a golden stag; but I alone could come anigh it. Then didst thou shoot and kill it even at my knees, so that sorrow was my portion and grief my fate for evermore. And in the stead of my golden stag gavest thou me a wolf-cub covered with my brother's blood."

And the Valkyrja answered gloomily:

" The rede of this is that thou shalt wed Sigurd, my betrothed, yet not by guile of thine. Guiltless shalt thou be, and he and I also. Yet this is our doom and naught can stay it, and great shall be the sorrow of us all. For he shall not live, though woe is me that I should be his death-dealer! and, with him dead, thou shalt be wife to my brother Atli. He shall slay thy brethren, and him shalt thou slay in turn. Thus the end of us all shall be woe and strife and the Twilight of the Gods."

" And is there no help? " asked Gudrun, with down-bent head.

" There is no help, since the Nornir have spoken," Brynhild replied; and, rising from the feet of Gudrun, she passed into her chamber, and all was sadness in the tower.

Then did Gudrun wend home to the Rhineland to wait for Sigurd and her fate; but Brynhild shut herself into her tower to work, in silks and gold, the Slaying of the Worm upon Sigurd's banner and, as the great coils grew and took shape under her fingers, so drew nearer the day of Sigurd's coming.

Then, after a winter in far lands, and the gaining of much fame, came the Golden Sigurd to Hlymdal to his betrothed. Sweet were the days of their love and life together, but all too few; for Brynhild, knowing the word of the Nornir, that soon he must pass to the land of the Franks even as it was decreed, bade him go forth to do mighty deeds, to help those in need, and to bear his great name scatheless as it had ever been.

So went forth Sigurd to his doom, and Brynhild, in bitter sorrow, hied her back to dreary Hindfell, there to await the fate she must needs not hinder.

Gudrun

GUDRUN

In the heart of the Rhineland lay the mighty city of Worms, home of the Niblung race. There in her rose-garden dwelt Gudrun, fair daughter of Gjuki, with her mother Grimhild, and her three brothers, Gunnar, the king, Guttorm, and Hogni.

Gunnar, the king, was powerful and rich, having hoards of gold and many brave warriors at his command; but chief of his treasures was his sister, Gudrun, the white-armed.

In quiet she walked one day, with her nurse, in the rose-garden beside the swirling river, when there came from the city a noise of great shouting.

" Go, nurse," she said, " and learn what this may mean. To me it seemeth a cry of joy."

The nurse went, and returned quickly saying:

" It is Sigurd, Fafnir's Bane, the golden hero of the Volsungs. Thy brethren ride forth to greet him at the northern gate. Come, nursling, that I may braid thy brown hair, and array thee in the gold of the Niblungs, for there will be feasting and welcome in the high-hall this night."

But Gudrun tarried, wistful, under the service trees of her garden with a foreboding of fate to come and of the end of her childhood's life.

Then throughout the Rhineland flew the word:

" The hero of the ages hath come; " and from

far and wide came folk to greet the dragon-slayer, master of Andvari's hoard, now returned once more to its Rhineland home.

In the high-hall, Gunnar, the king, held a feast, and near him sat his mother. Her bright witch-eyes looked upon Sigurd, and she pondered:

" If this man wed my daughter, naught should I have to fear for her from Atli and the wild kings of the East, any one of whom would wed her. He is troth-plighted to Brynhild, the Valkyrja, but that is naught. Have I not witch-lore to make him forget her? So also should we keep the golden hoard here in the Rhineland again."

So all that summer, while Sigurd hunted, played, rode, and waged war for the Niblungs did Grimhild wander among the mountains, brewing the magic draught of forgetfulness.

And the brethren loved Sigurd, and with all their lords was he in fellowship, save only with Hagen of Hunland, friend of Atli, whose deeds were evil, and who hated all that was brightest and best. So when the king prayed the hero to tarry throughout winter, he agreed, thinking: " In the spring will I fetch my Valkyrja maiden home."

But one autumn night, when all were weary with hunting and with the feast, came Grimhild bearing an ancient cup of gold to Sigurd, and, gazing with witch-eyes that faltered not, into the keenness of his eyes, said:

" In this cup I pledge thee, thou hero that shalt be my fourth son. Drink and see the desire of thy life."

And Sigurd looked straight at her with his

guileless glance and, taking the cup, drained it to the bottom.

Then fell a greyness upon his face, and all men were silent. He stood up and gazed around, unseeing; then, as one unmindful of his fellows, strode from the hall and was seen no more that night.

But Grimhild rejoiced, for she knew that her spell was strong.

In the morning, as Gudrun plucked service berries and late roses in her garden, there came to her the Volsung, as one in a dream. She was pale with the thought of his sorrow, though she knew not what had befallen him, and, letting fall her flowers, she held out to him her two white hands. Then he, seeing how fair she was, and Brynhild having passed from his mind, felt that with this maiden to love him, this strange nameless trouble of his mind would pass and all would be well; therefore took he her hands, saying:

"Gudrun, if troth may be plighted between us, here will I abide. But if, Daughter of the Niblungs, thou hast no love for me, then will I ride hence to-day. Say thou, shall I stay?"

And she, bending down her rose-flushed face, bade him stay, and he swore a mighty oath that never, while life was in him, would he forget her love.

So, hand in hand, they passed to the hall of the Niblungs, and a shout of joy went up from the chiefs of the land. And Gunnar swore blood-brotherhood with Sigurd, and they made haste to set forth the wedding feast; then did the crafty queen rejoice that all had fallen out according to her plan, and that the stolen Rhine-gold was once more safe in the Rhineland.

So Gudrun and Sigurd dwelt together in love, and the hero gave her to eat of the heart of Fafnir, and she, being of great soul, became nobler and wiser than all living women, save only the lost Brynhild. And then was born to them a son, who was called Sigmund.

And now did Grimhild, the plotter, turn her thoughts to that sad Valkyrja sitting bereft in her lonely tower at Hlymdal, broidering the deeds of her lost hero, and she said to Gunnar:

" Who so fit a wife for thee, my son, as Brynhild, daughter of Budli and sister of Atli of Hunland? "

And Gunnar, being willing, made ready to ride to Hlymdal, and Sigurd with him. Yet, ere they left, the witch queen called them unto her and taught them how each might take the other's shape. This seemed a thing of sport to the Volsung and, laughing his great laugh, he cried:

" Good mother and queen, wherefore do we need this witch-work? They say this Brynhild is now but a mortal maiden and needs but a mortal wooing."

" Thou knowest not what may befall," said the queen, " therefore heed well my runes."

So they went forth, Sigurd wotting little that these runes would bring him nearer to his doom, and rode merrily to Hlymdal.

And Heimar greeted them gladly and bade them tell their errand. Then Gunnar spoke:

" For the asking of Brynhild am I come. Thinkest

thou, Heimar, that she would wed with me and become queen of the Rhineland?"

"That can I not answer," quoth Heimar, "for no longer is she here, but back to Hindfell hath she fared. Strange and sad hath she been of late, but she holdeth fast to the rune of Odin, that only with the fire-rider will she wed—since wed she must."

"No fire fear I!" cried Gunnar.

"But thy horse," asked Heimar, "will he face fire? since ride must thou, even as ye would ride to Odin in Valhalla."

"That shall we see," cried Gunnar, doubting naught.

Sigurd the Volsung

THE WEDDING OF BRYNHILD

JOYOUSLY they rode over hill and dale until they came to the castle upon Hindfell, and round it still rose the quivering white flame.

For awhile they looked, then the mighty Gunnar, drawing his sword and shouting the war-cry of the Niblungs, rushed at the flame. But his horse, being afraid, swerved and turned and fled trembling back to the troop of men.

Then said Hogni the wise:

"Sigurd, lend thou me thy Grane, he feareth naught."

"That will I," said Sigurd, leaping to earth, "though I doubt me Grane will let none back him but myself."

And it was even so, for although King Gunnar mounted, no step would the grey horse stir. He stood like a rock in the pathway, save only that he turned his eyes to Sigurd as if to cry him shame in that he had let another back him.

Then Sigurd drew near and spoke low to the king:

"This must the queen, thy mother, have foreseen. So must I take thy shape and ride for thee, my brother in love, through the fire to woo thee Brynhild."

Then in the gloaming were the magic runes spoken, and Sigurd, in the likeness of Gunnar, sprang swift through the circling fire; and the fire died into grey ashes, and throughout the cold night did Gunnar and Hogni wait and watch for the Volsung.

But Sigurd strode through the silent halls until he reached the inner one, wherein, on the high-seat, sat the swan-maiden waiting, on her head a crown of gold, and on her white dress her arms lying listless.

Only her eyes of burning blue looked straight into those of the seeming Gunnar, and on her sad face was the woe of the hapless waiting for Sigurd, who came not.

And the seeker spake not, for his heart was cold with the weight of her sorrow, only he stood and gazed for a space.

Then the Valkyrja cried:

"Who art thou who cometh through the fire to disquiet me in my weariness?"

"I am Gunnar, King of the Niblungs, come to hold thee to thine oath and woo thee."

"Art thou indeed the first and best of men?" she asked, and her eyes sought his, in her heart-hunger for Sigurd.

"I am he," he answered, with bent head, "and thou, Valkyrja, shalt never be forsworn. This night must we be wed."

Then the swan-maid arose in her beauty and greeted the king, saying:

"An thou be the first of men, thou shalt be my king. Sit thou in my seat and take my troth-plight with this ring."

And she drew off the ring of Andvari and set

it upon his arm. So came the Fate back to Sigurd.

And when morning was come they parted, the kings to ride homewards, and Brynhild to go to Heimar's hall until the day of her journey to Worms.

Then was a mighty feast made, and great was the joy of Grimhild that her children were now mated with the best and most beautiful that the world held.

Many days did the great feast last, and of a sudden, at its ending, the mist rolled back from the spirit of Sigurd and he remembered the vows that he sware unto the Valkyrja. And he fled from that company in sorrow of heart, and knew not where he stood until he came through Gudrun's rose-garden to the Rhine bank. Dark and swift and sullen flowed the river, and Sigurd stretched out his hands and cried:

" Forsworn, dishonoured, I the Volsung! Thou curse of the ring, thou bane of the gold! Will naught be well until thou lie again in the arms of the waiting Rhine daughters? "

And he cast himself down amidst the flowers by the swishing black water, and so lay until the dawn. But when he arose his face shone with the golden light of peace; he spake unto no man of these things, but ever loved Gudrun the more, since no fault of hers was this sorrow; and he dwelt in friendship with Brynhild as a brother might.

But three winters went by, and Brynhild hated Gudrun each year the more, and pondered through long months how she might be revenged for the stealing of her love. No peace had she, but by night she wandered abroad calling on Sigurd, that should be her love for all the ages, and by day she sought, with bitter words and taunts, to humble Gudrun; but Gudrun, happy in the love of Sigurd, bore all and complained not.

Only her heart failed her in that she knew that Grimhild longed for the gold of Sigurd, and that Guttorm, her brother, joined her mother in this longing. Therefore kept she watch and ward lest hurt should come to her great husband, although she knew what was known to no other—that the blood of Fafnir, covering him, had made him safe from wounds, save only in the spot between his shoulders where had lodged the dead leaf; there alone might he be stricken.

Now, it chanced one day that the sister queens went to the bathing, and, as they went, Brynhild, well-nigh distraught with longing, flouted Gudrun even more than her wont, and the queen of the brown eyes grew wroth.

As she took place by Brynhild, the swan-queen cast on her a scornful glance and moved higher up the flood.

Then did Gudrun swim after her, asking: " Why shouldst thou shun me and move higher up? "

" Because thy place is below me," Brynhild cried, with a face of white wrath. " Thou, sister of Gunnar, as thou art—art but the wife of King Hjaalprek's thrall, a war-won slave. In days to come, when Gunnar and I sit with Odin in Valhalla, thou and thy slave husband shall wait without the gates. Could thy Sigurd have ridden the flaming fire?"

"Peace!" cried Gudrun, rising glorious in her anger, "who art thou that thou shouldst scorn the slayer of Fafnir and Regin, and many kings who wrought evil! Moreover"—she went on, in low, deep tones—"my Sigurd was it who rode the flaming fire on Grane and claimed thee for Gunnar. To Sigurd didst thou give the ring of Andvari, and here is it yet, set upon my arm."

She stretched forth her fair arm whereon shone the red-gold ring, and Brynhild went grey as a drift cloud. She cast one look of hate upon the Niblung queen, and, throwing her garments around her, sped to the wild wood and was seen no more that day.

But in the Mirk Wood found she Hagen, the evil one—a Hun and friend to Atli, her brother—who had been setting cunning traps for the wild things that run, and he said:

"What ails our golden queen? Can aught that I may do aid her?"

Then Brynhild burst forth with her hatred of the Volsung, crying:

"Help me with my vengeance!"

Then Hagen saw that through the queen he might perchance work evil unto the hero, whom he hated (as all wicked things hate what is good and brave and strong), and perchance compass his death, so that Atli might have Gudrun, whom he wished for to wife, and Guttorm the Golden Hoard that he so coveted, whereof part also might come to Hagen; so he said slowly:

The Two Queens

" This thing needs thought, but between us we may work it. Get thee home and show no anger at the tale."

But this Brynhild could not do. White and stony she rose up and lay down, speaking no word to the king nor to her maidens.

But after a while Hagen sent this message to her—that naught of her vengeance might be begun while she lay in her chamber alone, and he bade her come forth and be friends with Gudrun. So she came forth, heavy-eyed, and Gudrun met her with fair words, saying:

" Dear sister, let all be as if we had never striven by the Rhine. Flout no more, but let us dwell in peace."

So Brynhild, with the stone-cold heart and false lips, kissed the Volsung's wife and there was quiet.

But those two who hated Sigurd, and Guttorm who coveted his treasure, drew together and made a plot; but they told not Gunnar and Guttorm, for the kings were true to their oath of brotherhood, and would have slain the plotters rather than they should work Sigurd's woe.

Sigurd the Volsung

THE BETRAYING OF SIGURD

AND so the dreary autumn days sped on, and over the hearts of all lay a dim foreboding of evil at hand, and Brynhild, waxing thin, went heavily through the castle, white and still, with deep fire burning in her eyes.

To none was the evil known, save only to Sigurd through the sayings of Gripir the Wise and the death-word of Fafnir. Oft talked he with sweet Gudrun, as they sat in their chamber, where she loved to comb his red-gold locks—striving to prepare her for the sorrowful days in store.

But, though her heart believed, her mind would not, and she repeated ever that naught could harm him, since he had bathed in the dragon's blood.

" Thou mindest not that one spot," said Sigurd gravely.

" But it is so small a spot, dear heart, and none knows of it but I," and she laid her finger on that place.

Then Sigurd took her into his arms to kiss her, saying: " I would, my sweet, that it might be so."

But they wist not that a maiden of Brynhild was hidden behind the hangings and had noted all.

One day it fell out that the queens sat in their bower together. Gudrun and her maidens were broidering a banner for Sigurd, and swiftly the awe-

The Betraying of Sigurd

some grey coils of the dragon took shape under their fingers, growing apace as they laid on the gleaming red gold. But Brynhild, too sick at heart to work, sat idly by, tangling the silken skeins as the maidens laid them down by Gudrun, and scattering the ivory needles with restless, fluttering fingers.

Raising suddenly her heavy lids, she flashed a look upon Gudrun and said:

" Thy thrall husband is in peril."

Gudrun flushed red as she looked up but, speaking low, she only said:

" This insult is unworthy of thee, my sister; hast thou forgotten? "

" Forgotten? " cried she wildly, " will the gnawing worm of love and shame at my heart ever die? I tell thee, Sigurd is bound to perish by the plots of evil men, unless thou show me that one spot where he may be struck. If thou wilt do this I will bid my trustiest man-at-arms keep watch and ward over him."

This she said being set on by Hagen the Hun, for none knew, save Gudrun, where this spot lay.

" Sigurd holdeth watch and ward for himself," said Gudrun proudly; " natheless, sister, I thank thee for thy care."

Thus Brynhild failed, and with slow, dragging steps, passed from the bower. But there stole after her her dark-browed maiden, sister to Hagen, who, touching her, said:

" I, O queen, can show thee this spot; for, being one day within hearing, Gudrun and Sigurd spoke of this spot."

" It is well," said the queen, " mark thou the

place on the shirt that he shall wear to-morrow at the great hunt."

And on the morrow when all were gathered joyously in the castle-yard for the hunting of the boar, a small red mark was set between the shoulders of Sigurd, and Hagen and Brynhild laughed grimly, for now they knew that the hero was delivered into their hands for his undoing. But Sigurd held Gudrun long in his arms and kissed her, saying:

"Farewell, thou brave, true heart; bear thee well through the sorrow of heavy years to come, for in naught may we gainsay the Nornir. Yet shall we meet at last in Asgard, and our sorrow have an end."

Then Gudrun knew that nevermore should she speak with her love, but, daughter of kings, she bore herself bravely, looking steadfastly into his eyes as he turned away.

The hunt set forth, and she went to her bower, chill at heart, but Brynhild mocked her, crying:

"Thy hero will see fine sport to-day."

All through the day the hunt went on in the wild wood, and Hagen kept at Sigurd's back, biding his time to strike. But Gunnar, feeling something amiss, kept ever by his side also.

Then it chanced that, heated with the chase, they came to a running stream, and Sigurd leaped to earth to drink. As he stooped Gunnar came up, being also athirst, and Sigurd drew back that the king should drink first.

"Nay, brother Sigurd," quoth the great king, "drink thou with me as brethren should."

The Betraying of Sigurd

So they stooped and drank together, and the evil Hagen, stealing up behind, with one fell stroke of his spear on the small red spot, laid low the glory of the world, the Golden Sigurd.

Then rang through the wood two wild and terrible cries; the cry of King Gunnar for his brother foully slain, and the cry of Hagen, whom Grane had seized and bitten so that he died.

And the hunters came together in grief and pain and, raising the body of their hero, they laid it on a bed of spears and bore it back in gloom to the city.

And as they passed along in silence, a chill wind moaned through the pine tops, the robin ceased its autumn song; the ruddy leaves fell swift and thick from the beech trees; winter came in one breath over the land, and all things living mourned Sigurd dead, even as they had mourned Baldur the Shining God.

But Hagen, the traitor, was left a prey to the beasts of the Mirk Wood.

At two windows of the castle waited and watched the two queens.

And as the dreary train came in sight and she saw the bent heads of all, and Grane, riderless, behind the bier, Gudrun gave one shuddering cry of "Sigurd!" and fell senseless to the ground. But Brynhild caught a torch from the wall, and going down to the courtyard gazed on the face of the dead Sigurd with a laugh of triumph.

Then Gunnar spoke in anger:

"Woe unto thee, thou evil woman; get thee to thy chamber, and joy not that the light of the earth is quenched."

Sigurd the Volsung

But Brynhild spoke no word; she cast aside the torch, and, going to her chamber, laid herself with her face to the wall, and death was in that face.

And all through that night the Niblungs laboured and built a mighty pyre for the Volsung. By the will of Gudrun was it that it stood in the midst of her rose-garden, for she said: "What pleasure more shall I have, now that the light of my life is laid low?"

And at the dawn the bale was ready, and Gudrun kissed her love once more upon the mouth ere they lifted him thereon, and behold! at his side lay Grane dead. What use in life to Grane, wanting Sigurd? Could Sigurd ride to Valhalla wanting Grane? Him also did they lay with gentle hands upon the bale beside Gram, the gift of Odin, and at their feet two hawks.

Then, when all was lighted, forth came Brynhild, decked with gold and jewels, and bearing in her arms the body of young Sigmund, son of Sigurd and Gudrun, whom she had slain.

Terrible was she as she climbed upon the pyre and looked down on the face of the dead Sigurd in its peace and beauty. And with the weight of her weird her heart-strings snapped and she fell dead across the body of the hero she had loved and slain.

So passed Sigurd, hero of the ages, king of the true heart. But his name and his deeds passed not away, nor ever shall so long as the earth endures.

Gudrun's Woe

OF GUDRUN'S WOE AND THE RETURN OF THE RHINE-GOLD

BUT the sorrow of woe-worn Gudrun ended not here.

Bereft of husband and son she sat alone in her bower and bemoaned her for her love.

"Oh, for the life of the long-past days when Sigurd, my hero, who was as far above all other men as gold is above iron, lived and loved! No more can I dwell in these halls where my brethren begrudged me his true heart!"

And, wrapping a dark veil around her, she fled into the Mirk Wood, seeking sweet death and forgetfulness at the jaws of the howling wolves that abode therein. But they let her pass unscathed, and after many days she came to the hall of Thora of Denmark and there met with great welcome.

But, in the passing of time, news of her refuge came to Grimhild and Gunnar, and Grimhild said:

"It is but fitting that thou shouldst atone unto thy sister for the sorrow that thou hast brought upon her in the death of son and husband. Let us go seek her in noble fashion, with gifts of gold and precious stuffs—for with thee must I go, since ye have no skill in runes and fair words."

Thus set forth a mighty train of king, queen, and

many chieftains, and they came to the hall of the King of Denmark.

But Gudrun would have none of their fair words, but looked on them with sombre eyes. Thus had Grimhild foreseen, and she mixed for her daughter a drink in which all the magic might of the earth was mingled and gave it to her in a horn cut with runes of utmost power, so that Gudrun forgot the wrong done her by her brethren and her heart turned once more toward her kin.

Then Grimhild said:

" See, daughter, here art thou alone. Atli of Hunland, brother to Brynhild, would fain wed thee —a mighty king is he above all others. This do and thou wilt be the greatest queen throughout the world."

But Gudrun would not.

" My heart," said she, " is even with Sigurd, my one love."

Grimhild was angered and would not take her nay; then came a look of foreseeing into Gudrun's eyes, and she said:

" If this thing come about, evil from him will fall upon all my kin. Vengeance and death shall be their portion and that of Atli also. Mother, take this my warning."

Still did they urge her, and at last, wearied, she gave way.

" Small matter is it that I should suffer more, since the worst is overpast. Do ye as ye will."

Then that great company set forth with much rejoicing, and for four days journeyed through the forest; then did they sail for four other days on the Great River, and, lastly, ride for four days more ere they came to the high hall of King Atli, and there was the wedding set with much feasting.

But Gudrun smiled not upon Atli, but sat in chill silence.

Now after they had been some time wed, the king said to himself:

" Small comfort have I in this cold daughter of the Niblungs; more dower should I have had. The hoard of Sigurd should be mine, since its mistress is my wife. Yet never will those brethren give it up. I will bid them here and see what may befall."

Then he sent forth honourable messengers with Vingi, chief of his nobles, at the head. But Gudrun the queen, seeing much secret counsel, misdoubted her of their errand, and she knit round a gold ring a wolf's hair and cut upon a slip of wood runes of warning against Atli; these she gave into the hand of Vingi with word that he could give them unto Gunnar.

Now Vingi misdoubted him of those runes; he, therefore, carved over' them others, changing their sense so that they read otherwise and bade Gunnar come to the hall of Atli.

Well received was Vingi by the king and, after feasting, he delivered his message. Then spoke Gunnar aside with Hogni, the Wise, concerning this venture, and Hogni said:

" My mind misgives me of this gold ring bound with wolf's hair. It can but be our sister's meaning that Atli is of wolf-mind towards us."

" Nay," said Gunnar, " that cannot be, else would she not have sent these runes, bidding us to come."

And he handed the stick to Kostbera, the wife of Hogni, who sat by—a lady both fair and wise above other women.

She, looking closely, discovered the guile of that message, and said to Hogni:

" Canst thou not yet read runes aright, O husband? Beneath these are cut other runes that have been dealt with falsely, bearing the warning of Gudrun."

" Evil-minded are ye women," cried Hogni, " thankful am I that I think not evil when none is meant."

" How knowest thou that? " asked Kostbera, sadly. " Natheless, if thou wilt prove it, go; I warn thee that no friendship for the Niblungs lies in Atli's heart."

This much misgiving did the foreboding of Kostbera put into the brethren that, for surety's sake, they were minded to hide the treasure of Sigurd, Andvari's Hoard.

Thus, alone and in secret, they went by night bearing the gold and sank it in a certain place in the Rhine that they and the Rhine maidens, who guarded it, alone knew.

At dawn a great company went with the kings down to the ships, and with words of farewell they parted—each to his own fate.

To Vingi the lady Kostbera spoke:

" To me it seems that much evil will come of thy visit; would that it had never been! "

Gudrun's Woe

And the false Vingi made answer:

"May the high gallows and all evil things take me, lady, if I have lied in aught of my story."

In due time they came to King Atli's burg, and since the gate was barred, King Gunnar broke it down and they rode through. This angered Vingi, and he said, with an evil look:

"Now will I seek thy gallows-tree! Softly have I spoken hitherto to lure ye here, but this was my intent—to cause your deaths."

Hogni laughed.

"No fear have we of gallows-tree nor yet of armed men," he said, "but swift punishment do we mete out to traitors."

So saying, he felled Vingi with his axe so that he died.

Passing on, the Niblungs found King Atli and his chieftains drawn up in battle array and he rode forward to have speech with them.

"Have ye come," he asked, "to deliver up the gold that is mine of right—since it is Gudrun's?"

"Nay," answered Gunnar, "never shalt thou have it."

"Long have I been minded to take that gold and, with it, vengeance on ye all for that deed of shame, the death of Sigurd, Fafnir's Bane."

Now, as they parleyed, came word to Gudrun of her brothers' coming and, running forth, she went to them and kissed them on the mouth saying:

"Why did ye not heed my runes? Yet to each must come his fate."

Then she drew back into her bower and the

battle set on. Twice rode Gunnar and Hogni through the hosts of Atli, driving them back with exceeding fierceness, as leaves drive before the autumn wind, until they reached even into the great hall. There was Gunnar overborne and laid in fetters, while Hogni of the Stout Heart felled twenty to his own hand ere he, too, was borne down and set in a place apart from his brother.

Then came Atli, as they lay bound, and said to Gunnar:

" If thou wouldst have life and breath, tell me the hiding-place of the gold."

For long did Gunnar ponder, then made answer:

" Show me first the bleeding heart of Hogni, my brother."

Now Atli was loth to slay Hogni in this wise and he bade them kill a thrall, a swine-keeper. This they did and, cutting out his heart, bore it to the king; but Gunnar said:

" Trembling heart, that trembled in life as it does in death—no heart of Hogni art thou."

Then they fell, by Atli's command, upon Hogni and truly took his brave heart to show to Gunnar.

The king looked fixedly upon it and said:

" The mighty heart of Hogni is it. Steadfastly does it beat in death as in life."

Then he laughed out a great laugh of scorn:

" Now, O Atli, I alone am left who know the dwelling of the gold. Never more shall any own it save the Rhine maidens from whom it came."

And again he laughed so that Atli, enraged,

bade men cast him into the serpent pit, and this was done. But the serpents made not an end of him, for Gudrun came secretly and cast in to him a harp, whereon he smote with such skill that the evil worms were soothed and slept—save one only that crept close and smote him to the heart so that he died.

Then, with woe for the loss of her brethren, became the heart of Gudrun all evil, yet she spoke fair words to Atli and prayed that a mighty funeral feast should be made for her kinsfolk and his. And this was done, though gloomy she sat and proud, through it all.

But that night, when Atli had drunk much and lay asleep upon his bed, she came into his chamber and thrust him through with a sword so that he died—but not until he had heard all her words of bitterness.

"False words hast thou spoken to me; false deeds hast thou done to my kin. Full of strife and battle has been my life in thy halls; gone is the peace that I had with Sigurd, my beloved. Better had I remained his widow for my life long than become the wife of hated Atli."

"Hate if thou wilt," said the dying king, "but see thou that I have a goodly funeral pyre."

Thus did Gudrun the queen, for casting fire into the great hall, she made the whole burg of Atli his burial fire, so that the flames soared to high heaven to cry out the death of Atli, the great lord of the Huns.

Thus ended the race of the Niblungs, lords of

the Burgundian land, and thus, once more, came the Golden Hoard back to rest in the arms of the Rhine maidens, its warders.

And even unto this day, at whiles, may be heard their sweet songs of joy, as they float at sundown, watching over the treasure of Sigurd the Volsung.

VÖLUND THE SMITH

ONCE, long ago, there lived in Finmark an elf king who had three sons, Völund, Slagfidr, and Egil, to whom sport and the chase were the best things in life. Skilful were they in wrestling, swimming, and running on their snow-shoes or skates of bone, which they named ice-legs. But most of all did they love the chase, and to the end that they might enjoy it more, they built themselves a hut in Ulfdale, on the shore of the Wolf's Water, and here did they dwell for the most part.

Now Völund, being somewhat lame, could not always go with his brethren on the longer winter chase; therefore, at times would he stay in the hut and fashion wondrous jewels of gold and bright stones, as he had been taught by the mountain dwarfs of Finmark, who, for his father's sake, held him in great friendship. And his skill was lauded in all the lands where the Norsemen sailed, and even down to Miklagard[1] on the shores of the Middle Sea. Yet would Völund never work for gold, but only for love of beautiful fashionings and for friendship, so that his much-sought work was hard to come by.

Now, it fell one day in the soft spring time, when the larch tassels tossed their red and green in the whispering breeze, and Baldur's-Eye crept out,

[1] Constantinople.

sweet-blue, to meet the sun, that Völund came back from many days' wanderings among the bare black mountains. Glad at heart was he, for the berg-trolls had sent him word of a marvellous jewel of green and blue and purple and wondrous burning fire that lay hidden in crevices of dull brown rock, and after much searching he had found it. In his leathern pouch it lay, pebbles like unto naught but the rainbow bridge of Asgard that is guarded by Heimdal of the golden teeth; and Völund's mind was full of thought of how he should use the stones. As he passed through the forest, near to Wolf's Water, he heard the voices of his brothers returning from the hunt, and together they made their way towards the hut, talking as they went. But as the path got clearer and the light showed through the edge-most trees, the brethren stayed their steps and looked one at another, for there was sound of voices and laughter from the water-side, and, walking stealthily, they peered forth and beheld three maidens sitting on the golden sand in the morning sunlight. They had bathed, for their feet were bare and their hair fell round them unbound, and beside them lay three dresses of swans' feathers of dazzling white.

Then the brethren knew that these must be the Valkyrjar, Odin's choosers of the slain, and princesses, and they went forth to speak with them. The maidens sat spinning flax and looked up, smiling, as the king's sons drew near.

" A fair morrow to ye, hunters," said the first, whose hair was black as Odin's ravens, " and good sport. I am Hladgrun, daughter of King Hlödvir,

Völund the Smith

and she with brown locks is my sister Alvit; we are the foster-sisters of golden-haired Olrun, who is the daughter of King Kar, our father's friend. From All-Father got we leave to fly to Wolf's Water, for the fame of its golden sand and deep blue water has travelled far, and here would we bide, at least for a while."

Then the brethren bade them welcome, and made them couches of their finest skins, and the maidens abode there until it fell out that Egil married Olrun, and Slagfidr Hladgrun, and Völund chose Alvit of the soft brown hair.

Great was their weal during many years, for the warrior women followed the chase with their husbands, and when Völund abode at home, Alvit stayed with him and helped in the welding of rings, joying in the blending of the marvellous stones.

Now, when seven years were forebye, there came a shadow over the homes by Wolf's Water. The Valkyrjar grew pale and still; in the eighth year drew they the swan-feather dresses forth from the great chests where they were hidden, and preened them on the yellow sands in the sun.

To Egil and Slagfidr this became a jest.

" Wives, will ye fly away as ye came? " they asked; " and shall we need to seek ye in Asgard? "

Only to Völund was it earnest, as he worked and thought; and he said one day, looking deep into Alvit's eyes:

" Wife of mine, is there aught in thy mind that thou hast not told me? "

And she answered, sighing:

Völund the Smith

" Völund, ever is it the Valkyrjar's weird that they must go when Odin calls."

" And wilt thou return no more? Wilt thou forget? "

" Never shall I forget; yet of my return I may not speak, since the future is hidden from me. But, Völund, All-Father is merciful and kind, and of a surety we may hope."

So Völund went about his work and made no sign, nor said he aught.

One winter's day, in the ninth year, when he and his brethren returned from hunting there was no answer to their call; the huts were empty and the swan-feather garments gone.

Then told he all unto the others:

" With Valkyrja were we wed, therefore must we suffer; for to Odin do they first owe fealty, and who are we that we should contend against the gods? "

" Odin has many choosers of the slain," said Egil gloomily; " we each had but our chosen one. Surely this is scant justice? "

" Justice or not," cried Slagfidr, " I go to seek Hladgrun, even to the foot of Odin's seat."

" And I with you," said Egil.

" Such journeyings as thine are not for me, my brethren, since I should but hinder ye in your going. Therefore, here will I abide, to keep the home and welcome them should they return. Here is gold for your plenishing and weapons of the best, that I have wrought. Go forth in peace, and come not back alone."

So the two did on their snow-shoes, and clad

themselves in their warmest skin-coats, to set forth on their long travels, but Völund stood by the hut door to watch, as Egil turned towards the east and Slagfidr towards the south; and he watched until the crisp noise of their footfalls on the snow died away in the forest; then he turned to his lonely work and sighed.

So he sat, always by the open door, making precious rings of gold for Alvit, always for Alvit; and when fifty were finished he strung them upon a thread of grass, until they numbered seven hundred in all, and no two were alike.

Now it came to pass that Nithudr of Sweden, a wicked king, who worked but evil in the land, heard of the fame of the Smith's gold work, and sent unto him saying:

" Give me of thy work, rings and a necklace and a beaker, and for these shalt thou have much gold."

But Völund sent back word:

" Gold have I in plenty; and the work of my brain and hand is not for thee."

Then was the king angered; but his evil wife, who desired these things greatly, said:

" Speak him fair, O Nithudr, so may we yet come at them."

And Nithudr sent again, saying:

" Choice furs have I, such as are sought after vainly even by the kings of the south. From my hoard shalt thou choose all that thou wilt, so that thou give me but my wish."

" Furs have I in greater store than Nithudr,"

said Völund, " both black and grey, and white and sable. With these can he not tempt me. Leave me in peace."

When the messengers brought back this word, Nithudr was so wroth that all trembled, and he bade armed men go to Wolf's Water and take Völund and his treasure and bring them before him.

At midday came the men to the hut, and found all still and the door wide open, as Völund was wont to leave it lest Alvit should return. And the light shone on their mail coats as they dismounted by the hall-gable and entered, and, seeing the rows of shining rings hanging by the wall on their threads of twisted grass, they took one, then hid in the forest until the Smith should return.

Woe unto Völund! no bergtroll came, friendly-wise, to warn him of the foe so close; no Valkyrja wife called to him out of the fleecy white clouds to shun the home-hut, and at eve came he back, dreaming of Alvit.

As always, after an absence, he counted his rings, and fast did his heart beat when he made the tale but six hundred and ninety-nine. Once and again he went through them, and he said:

" It must even be that Alvit is here; for sport hath she taken the ring and hidden herself."

In haste did he light a fire of crackling fir-cones, for the night was chill, that the ruddy gleam from the open door might lure her home; and oft did he stand without to pierce the gloom; but all was dark and silent, and to his cry of " Alvit! " came no answer.

Völund the Smith

Then set he bear's flesh to roast—a portion for two—and high blazed his faggots of wind-dried fir-wood. Then he sat him on the rug to count once more the rings; and as he sat heavy sleep came upon him, and he fell back dreaming of Alvit.

And the men of Sweden came creeping forth, their shields shining in the cold moonlight, and fear was in their hearts, for some said that Völund knew the magic of the runes, and could weave them as he would; and they took him sleeping, set his legs in iron fetters, so that when he came to himself he was prisoner, and was carried bound to Nithudr.

The wicked king rejoiced greatly that a man so dowered should be prisoner to his hand, and bade the men set Völund before him. Weary and heart-sick, the Smith sat upon the ground before the king's seat, for, by reason of the heavy fetters and his lameness, and the long journey from Wolf's Dale, could he no more keep upright.

And Nithudr taunted him with bitter words.

" Thief! have I got thee at last? " he cried. " How long have I borne that thou shouldst come unawares, under seeming of the chase, and steal my gold! "

But Völund sat with bent head, and made no sign.

Then shouted Nithudr:

" Whence didst thou filch those untold treasures? The gold is gone from Gnita Heath back to the Rhine and the hills of Rhine are far away, O Lord of the Elves! "

Still Völund answered not.

Völund the Smith

Once and again the king mocked him, and at length he spoke:

"Who art thou," he said, "that darest to fetter a king's son!"

"When a king's son is but a common robber he meets but a robber's fate," said Nithudr.

"By help of my troll-friends is my gold found," said Völund, "and none of it is gained in thy land, O king; therefore let me go back to Wolf-Dale."

"Never shalt thou win back until thou hast given me my desire. A ring I have, since my thralls brought it to me, thy sword shall I take now, since it is the work of Völund the Smith, and none there is like unto it; but I still lack the necklace and beaker. Give me these and thou goest free."

But Völund answered sullenly:

"Kill me if thou wilt, but my work is for my friends, and thou art my bitter enemy."

Then knew not Nithudr what to do, and he was minded to let Völund go, for he feared the King of Finmark and the bergtrolls, who loved Völund.

But the wicked queen would not, for above all she desired a necklace of the rainbow stones of which the fame had gone abroad, and she said:

"It is well for thee, Nithudr, for thou hast the Smith's mighty sword, and to Bodvild our daughter hast thou given the ring—but what have I? It seemeth to me that he is a man of craft, as are all forest-dwellers, and so long as thou hast his sword will there be no peace with him in our steading. Lame him still further, so that he can

Völund the Smith

never flee away, and put him on the Island of the Salt Farm, then can he work all that thou wilt."

And this shameful thing was done as she had said, and Völund was set upon the Island alone, with gold of the Swedish king; and, because of his loneliness and misery, he wrought for Nithudr many wondrous things, yet over all murmured he runes that ill should befall all who owned these treasures. At times there came upon him mighty wrath and despair, so that he smote upon his anvil with such force that it crumbled as if it had been clay; then must he weld another anvil ere he could work again, and as he welded he sang:

> " Alack! for the sword, my companion,
> Alack! for the steel I forged and ground;
> Now it has passed to an alien,
> My faithful friend hangs at Nithudr's belt.
> Lost to me is its brightness,
> Nevermore will the runes on its blade
> Whisper to me of their magic!
> Alack! for the ring of my fashioning,
> Alack! for the glory of Alvit;
> To an earth-maid hath it been given,
> My Valkyr is lost, is lost as my life,
> For no end is there now to my sorrow! "

So each day did he dream more of vengeance, and each day became he more gloomy and sullen, until, after many weeks, there came this hope of revenge, although for long was it delayed.

Two young sons had Nithudr, brethren of Bodvild, the princess, who were like to their parents for greed and cunning and hatred of all things good. And it fell out one day that the boys, looking out over the sea, spoke together of the Island and the prisoner thereon.

" Bodvild hath said that she weeps for him and

his dreary fate. She hath begged our father to let her go and visit him," said the younger.

" Aye," said the elder, " but she is a fool. Our mother saith that he is evil and good for naught but to make us jewels and swords. Could we not go over and seize his treasure ere it come to our father and mother? So should we be rich for ourselves and not beholden to the king and queen. For our mother saith he is a constant danger, and when she hath gathered enough treasure from him she will send thralls to slay him."

" And Bodvild crieth shame upon her, and our mother, being angry that Bodvild will not give up the ring of Alvit, hath put her away in the inmost room of the house."

" That matters naught to us," said the elder; " gold must we have if we would be powerful, and here is gold for the taking, and only a lame man's life between."

Whereby it will be seen that evil parents have evil children, and Bodvild alone was mild and merciful.

Then the youths sought cunningly how they might get a boat to cross to the Island of the Salt Farm, yet must they be very wary since Nithudr and the queen doubted of every man—as is the way of the wicked—and ever kept strict watch on the outgoings of all.

But they knew not that Egil, having wandered through the world seeking the Valkyrjar, had returned and made his way by night to the Island, where Völund received him gladly, and hid him every day at the coming of Nithudr. And first he asked:

Völund the Smith

"Comest thou, O brother, with news of my wife?"

"With news, O Völund, but naught of good. Alvit and Hladgrun and Olrun saw I, but it availed nothing, for Odin is wroth because of their long absence and hath said that, an they will to return and dwell with us, they shall become but mortal women, to dree the weird of mortals, in sorrow and suffering and death. And they have taken counsel together, and their word is that this they could not thole. Life to them is it to ride the clouds and bear the chosen to Valhalla, rather than to suffer the love of us mortal men, and die with us. Slagfidr have I seen and told, and he would that I should sail a-viking with him, but I could not, for thou wert alone and lame, so that I must seek thee according to my promise. He would that we should together seek him in Miklagard where thy gold-work would be much sought."

But Völund shook his head and said:

"With Alvit went the light of my life. If I leave here—and fain would I do it—back to Wolf's Dale shall I go, there to abide always. But first must I have vengeance on Nithudr and his wife, in that, by their act, nevermore can I hunt in the forest, and seek for precious stones in their hidden homes."

To this did Egil agree, but never could they compass a plan for the undoing of Nithudr. And Egil, for employment and seeing that nevermore might his brother win from the Island unless by flying, bestirred himself to gather swans' feathers wherewith he might weave a dress for Völund, like unto those of the Valkyrjar. And the white swans, gathering round at eventide, brought him more feathers, and sang their sweet strange songs to hearten him at his work; and so two winters went by and the Smith was of good heart, for the king knew naught of Egil's presence, since by day he hid in a sand cave on the far side of the Island.

At the end of this time it befell that Bodvild, who dreamed ever of the lonely worker of the Salt Farm, broke the precious ring of Alvit, and, since none could mend it, and she dared not tell her father, she took four thralls and her two maidens at early dawn, while her parents slept, and rowed over to Völund.

And when Völund beheld the fair maiden drawing nigh alone, he went forth to meet her and give her greeting. " I am Bodvild, the king's daughter. O Smith, I come to beg that thou wilt mend my ring," and she showed the ring of Alvit, lying in two pieces in her hand, and Völund, thinking of his false Valkyrja, looked so long and hard upon the princess that her face flushed, and she dropped her bright head and waited. Then Völund spoke:

" Thy ring will I mend, O Bodvild, but only at the price of thy love."

" My love is thine," she answered simply, " and has been since the long-past day when, lame and despairing, thou wert brought before my father."

" And if I wed thee, will he be wroth? " asked Völund.

" So wroth that I doubt not he will kill me, because of my mother's hate of thee. But what matters it if I have thy love? "

Völund the Smith

"Spoken like a king's daughter," said Völund. "Call in thy maidens and thralls that we may plight our troth."

And he called unto Egil, his brother, that he also should witness; and there, before the seven in the dark smithy, were Völund and Bodvild wed. And he set upon her neck a great golden collar, set with glittering stones. "For," said he, "since rings were made for Alvit, thou shalt have none from me, but arm-rings and necklaces, and girdles and crowns—gold for thy golden hair—as many as thou wilt."

Then Bodvild kissed him on the mouth and went over the sea to her home, but oft at early morn thereafter she sped across to spend what time she might with her husband.

And now came it that Völund often laughed and sang runes over his work, so that Egil said:

"Hast thou a secret joy, O brother, that thou sittest no more in gloom and silence?"

"A joy have I," said Völund, "in that my vengeance draweth nigh. Is thy swan coat finished?"

"In three days will it be ready; but, brother, thou wouldst not hurt Bodvild?"

"Nay, she hath been the first part of my revenge, in that Nithudr would rather that she lay dead than that she should wed me. She is a gentle maid, and will dwell with me at Wolf's Water. Soon will come the viper's spawn, his sons, and my work will be done."

And so it fell out, for, by constant watching, the youths in the end made their way to the smithy unseen by the king or the queen, and strode in upon Völund as he worked. His great chests stood open, and their greedy eyes beheld the jewels that lay heaped therein.

" Give us of your rings and gold," they cried roughly.

" Come hither alone to-morrow," said Völund, " and see to it that ye tell none of your errand—neither maidens nor hall-men. Then shall ye take and carry off all that ye will."

And he hammered the more grimly on his anvil.

Then the lads departed, saying low:

" And what shall hinder us from killing that lame man and taking all? "

While each in his heart thought:

" Then will I slay my brother so that the hoard may be mine."

Hustling each other in their haste and greed, they came next day ere the dawn, and running to the chest, struggled which should seize the most. But as they knelt and fought, their heads being within the chest, behold the iron lid came down upon them and cut off their heads.

Then was Völund's revenge fulfilled. He took the skulls of the king's sons and set them in silver as a gift for Nithudr; their eyes and teeth by his runes he changed to stone and set as jewels for the queen.

So came the youths home no more, and therefore had the wicked queen no rest: ever did she wander by the shore and in the birch-woods seeking her sons, who came not; while the king sat in his high-seat waiting gloomily, and Bodvild kept her chamber, and so the days went by.

And one day his wife came to Nithudr and said:

" Wakest thou, Nithudr, King of Sweden? "

" I wake ever," he answered, " for joy hath fled and no more can I sleep by reason of the evil counsel that thou gavest me; for I fear me that by this it is that our young sons have come to their death. I would fain speak with Völund, for it is borne in upon me, that by reason of my cruelty to him has this sorrow come upon me."

" Völund is here," came a voice from above, and going to the door of the high-hall the king saw Völund, clad in the swan-feather dress and holding in his arms Bodvild the princess, high above him in the clouds.

Then the king called aloud:

" Tell me, thou master of runes, hast thou seen my sons? "

" Swear unto me first, by point of sword, by Sleipnir's mane, by ship of Odin, and by Urda's fountain, that thou wilt never harm my wife, no matter what her name, nor do hurt to child of mine."

And Nithudr swore by all these things.

" Then go to the smithy, the prison where thou didst set me, and under the dust in the pit beneath the bellows wilt thou find thy sons. From their skulls hast thou drunk thy mead, round thy queen's neck hang their teeth and eyes."

Then the queen shrieked aloud and tore from her neck the fated stones, and the king cried:

" Would that I could take vengeance on thee,

Völund the Smith

O Völund, and on my daughter, but for my oath's sake I cannot. Neither could aught, save Odin's ravens, tear thee down, nor could the most cunning archer reach thee in thy clouds. Go with Bodvild, and trouble me no more."

So Völund, laughing and bearing Bodvild, soared away across mountains and forests and tarns to his loved Wolf-Dale; and there they dwelt until their deaths, and they had a son named Vidrek, who became a great hero in after times in the southern lands. And ofttimes came Slagfidr and Egil to talk with them, and show them of their booty; and through their tales was it that the little Vidrek was minded to go forth in search of adventures.

His story is of a later time and cometh not into this place, but the fame of Völund the Smith went forth through all lands so that after many hundred years in far-off countries, even England, did folk still call upon him, when in straits, to do their smith-work.

RAGNAR LODBROG

In Viking days there dwelt in Gothland a mighty jarl who was called Hraud. No sons had he and no child save one daughter, who was so beautiful that seldom was she called by her given name, Thora, but was known to all as Borgar-hjort, which means Hind of the Castle. By her noble birth and lovely face was she fitted to be chosen by Odin for one of his Valkyrja; but she would not, for her nature was gentle, and she feared and hated strife.

Then, since her father must be oft away on Viking raids, and since he was much disquieted at leaving her alone, did Jarl Hraud give to his daughter a magic box wherein lay much gold, and upon the heap of gold a small dragon.

And Thora, opening the box, cried out in wonder at the strange worm:

"Why should I keep this laidly thing, my father? Surely it were better away in its haunts on the far inland wastes?"

"Nay, daughter mine, this is a witch-worm, and I give it thee as a guard. Since I must leave thee for so long, it is meet that thou shouldst have a better ward than thy faithful servants. This dragon will grow to be the fear of all the lands around us, so shalt thou dwell in peace until I come again."

So Jarl Hraud hied him forth in the long days of spring, and Thora dwelt alone in her castle. But daily grew the gold, and with it grew the dragon, until he became too great to bide longer within the castle; and Thora took counsel with her nurse and the overman of the jarl what she should do, and the overman said:

"Since the worm is here to guard thee, Lady Thora, he must not be driven forth, else will thy father be angered. Were it not well that he should bide without the castle, and so fright all that come with ill-intent?"

And the nurse said:

"Even so might it be, O nurseling; I will go speak to the worm."

Then she spake fair words to the dragon, and he saw the reason of her speech, and dragged his slow length without the castle, until his coils encompassed it on all sides. And so he lay that none could go out or in without his knowledge and sufferance, and Thora was well guarded. But as time went on the worm grew evil and would let none pass save those who brought in food; each day became he greater in strength and venom, and when the short days came, and the jarl returned to winter quarters, he was kept without his house, while Thora lay within, and might no more come forth to greet her father.

Then was Hraud in great straits, and he went unto other jarls, his friends, to take counsel what should be done. And all said one thing:

Thora

" Kill the dragon."

" Easy is it to say," quoth Hraud, " but who shall do this deed? Too old am I to fight this monster; moreover, a witch-worm is he, and more to be feared than all worms save Fafnir, guard of Andvari's hoard."

" This do," said an aged jarl. " Cause it to be told throughout the north that whoso kills the dragon shall wed Borgar-hjort. So shall she have a worthy mate and be the fairest, best, and richest of the maidens of the land."

And all liked this counsel well, so Jarl Hraud sent abroad the word, and there was much talk and stir in many lands.

But most was there stir in the heart of Prince Ragnar, son of King Sigurd of Sweden. Oft had he heard of Thora, and his mind was filled with the thought of the fair maiden, dragon-warded. And he asked much of the messengers concerning the worm and his ways, and he caused to be made five cloaks of coarse wool and five pairs of breeches, and these he had boiled in pitch so that they were hard and like unto garments of thick leather, but some men say that they were but wild goat-skin. Be that as it may, from these breeches got he his name of Lodbrog, which means leathern breeches.

So Ragnar went up against the dragon, and after a mighty fight, wherein the great beast sought to poison him by biting through his clothes but was unable by reason of their thickness, he struck his spear so forcefully through the back of the dragon that he was unable to draw it forth, and the shaft broke off and remained in his hand. And the dragon cried aloud in its death-pain:

"Ah, that I, the terror of the nations, the warder of Borgar-hjort, should be done to death by the guile of a stripling. Tell me, youth, how many winters hast thou?"

"But fifteen," answered Ragnar straightly.

"Thora! Thora!" cried the worm, "this fifteen years' boy hath ventured much for thee. Take him, love him well, for he will cherish thee greatly."

So died the dragon, and this is the true story, though some say that the dragon was but a chief named Orm, set over Thora's castle by Hraud, and who, on his return, would by no means give her up. But this is false, since no word is there at this time of a Jarl Orm in Gothland.

Then went Prince Ragnar unto the jarl, where he sat awaiting the issue of the fight.

"The worm is dead, O jarl," said Ragnar, "and in proof that I have slain him here is my spear shaft. The point is set in the worm's back where ye may find it. Now claim I thy Borgar-hjort for my bride, to love her ever for her goodness and her grace and her beauty."

"And blithely shalt thou have her," quoth the jarl, and led the youth into the castle, where Thora straightway loved him and plighted him her troth.

So they loved and lived in happiness many years, and two sons were theirs, Ragnvald and

Thora

Agnar, who went a-viking with their father, until it fell out that gentle Thora died and Ragnar thereafter could no longer suffer his home. Putting his realm into the hands of first one, then the other of his sons and of his wise counsellors, he, with the other son—each in turn—sailed a-raiding. It is told of him that he sailed east, even up the Vistula, and southward by the great rivers of Gardar, unto Miklagard,[1] to visit the Norse Varanger guard of the ruler there; on the Danube slew he eight chiefs and gathered much spoil; and it is written that he returned by the Middle Sea. In all lands was he known and feared, yet could he not always conquer, for, as time went on, first Ragnvald and then Agnar fell in battle, and Ragnar was indeed alone.

Then went he no more to Sweden, but sailed, ever plundering. Throughout Iceland, the Isles of the West, and the Fleming's land was he known and dreaded in the summer-time, and the winters passed he in the warm havens of the Middle Sea. But his men grew hungry to see their homes, and would fain return northward.

" Seven years have we followed thee unmurmuring, O Ragnar," they said, " now is it our turn that we should sight the shores of Sweden."

And Ragnar knew that they were right, and he said:

" Black were my locks when I came forth with ye all, grey are they now with my unending sorrow. But it is not meet that all should suffer for one, therefore, hoist thee the great sail and let us return

[1] Byzantium.

to our own land. It may be that so I may find comfort."

Then was great joy through all the long-ships, and their beaks were speedily turned northwards so that, with fair winds, they swiftly sped towards their home. But ere they came into Sweden much befell, whereof the tale must be told.

Aslaug

ASLAUG

Now it is told that Heimar of Hlymdal, brother to Brynhild the Valkyrja, had, at fostering, a beautiful woman-child, named Aslaug. None knew her race, but most thought her to be the child of Sigurd and Brynhild. Be that as it may, at the death of Sigurd, fearing the vengeance of Gudrun for her slain son, Sigmund (whom Brynhild had killed and laid upon the bale of Sigurd), Heimar caused to be made a great harp, with a golden stem, wherein he hid the maid, with rich treasure of gold and jewels, and onions for her to eat, since these give strength and sustain life long. And he dight upon him the clothes of a wandering skald, and, the golden harp upon his back, fared forth to seek safety for Aslaug. Through many lands went he, letting the child out from her hiding-place to run when they were hidden from the eyes of men, and playing on the harp to comfort her when she sorrowed for her home.

Now, it befell that late at eventide on a dreary day of rain, Heimar came to a lonely place in Norway, that was called Spangarhede, but now is it called Krakebeck, or Guldvig, because of the king's daughter, who lay hidden in the golden harp.

There dwelt an old man Aki, with his wife Grima, and Heimar, being wet and weary, smote hard upon their door. Now Aki was absent, and

Ragnar Lodbrog

Grima was long in opening, for she would first look well through a crack in her wall to know who this stranger might be.

And seeing this man of kingly height and noble face, with the golden harp upon his back, she unlatched the door, and asked:

"What wouldst thou in our poor house at night?"

"Shelter would I have from the rain, good mother," answered Heimar, "mayhap, fire to dry my clothes, and food, for I am wet and weary."

"Shelter canst thou have," said Grima, "but neither fire nor food. Few peats have I, and what I have must wait for my man, who journeys far to-day."

"Nay, but if I pay thee well canst thou not give me aught?" asked the king.

"Take then," said Grima, "there are peats; kindle fire thyself."

And as Heimar busied himself with the fire, the wicked hag sat glowering, and she noticed that as he stretched forth his arms to the blaze, there glinted under the fringe of his harper's frock, the shine of a great gold arm-ring, and she thought:

"None know that this stranger is here. Good were it if we could take his gold, for weary am I of being poor."

Then came a knocking at the door, and old Aki entered, bearing peats upon his back. Him did Heimar greet in friendliness, and together they ate of the rye-bread that Grima set before them, and talked as they ate:

"Always poor have I been," said Aki, "and

oft an hungered. Fain would I give over work and take mine ease."

"That perchance may come to thee soon, friend," said Heimar cheerily, as he laid him on the settle to sleep, and he thought to himself:

"Perchance when the old pair are quiet in the byre, may I draw forth Aslaug to sleep here by me."

But Aki and Grima slept not, and so long did they keep moving that Heimar fell into that sleep that was his last.

For Aki, set on by the wicked Grima, stole in and killed him as he lay. So died the noble Heimar for the sake of his sister's child.

Then through the mirk dark night they bore the body forth, and buried it deep in the sand-dunes, and set stones a-top, and, as the late dawn came, crept back to the lonely hut.

There Grima laid hold upon the harp, and the strings wailed mournfully, so that she pushed it from her in haste. In falling the pillar burst open, and there lay Aslaug, the maiden, smiling with the wondrous steel-blue eyes of Sigurd the Volsung. In terror the two fell upon the ground, and Grima, shuddering, cried:

"Kill her! kill her! O Aki! lest through her words our doom come upon us."

"No more killing will I do," said Aki gloomily. "Our doom is here and our own guilt will never die, since I slew a man by stealth and not in fair fight."

But Aslaug spoke never a word. She busied herself in gathering up the gold and jewels that were scattered over the ground, and in folding the golden stuffs and laying them again within the harp pillar.

" Tell me thy name," said Grima at last.

But Aslaug shook her golden head and smiled once more; and still she said no word, so that the two believed her dumb.

Now it had fallen out in this wise:

When Heimar went forth from Hlymdal he feared greatly lest the child should babble and tell her name and parentage, therefore had he straightly bidden her to speak to none but himself, and that only when they were alone.

Now Heimar being gone, Aslaug spoke no more to any living being.

Then, since Aki would by no means kill the child, Grima took her and darkened her white skin with juices of the bracken, and hid her golden hair under a rude cap of wadmal, so that none might know in her the princess in silken garments, and sent her forth to tend swine in the forest and goats upon the seashore. These they had bought with a part of Heimar's treasure; the rest they hid in safe places, and the golden harp they destroyed. No longer Aslaug was she called, but Krake.

Dreary was her life and strange were the thoughts that came to her as she sat alone among the pine trees, whispering all the words she knew lest she should forget the speech of men. So many and far-reaching were these thoughts that she grew wiser than others of her kind, seeing that she—like Sigurd, her father—knew even the speech of birds.

At whiles, when storms raged on the sea, she

would sit upon the sand-dunes and sing. And so strong and beautiful was her voice that the sailors far out from shore said:

" Hearken, how the Valkyrja ride the storm."

One strange thing did she, nor ever knew the reason for the doing, since the birds told her not.

Upon the shore, among the dunes, was a lonely barrow with a few stones thereon, and Aslaug said to herself:

" Perchance a dead man lies here; a viking who should have been sent to sea in his burning ship. Be that as it may, a mighty barrow shall he have. Some man of power and might must it be, since by night the flame flickers ever thereon."[1]

And daily went she to the shore, bearing ten stones that she set upon the barrow.

And the old people, watching her, trembled and said one to the other:

" She knows that the harper lies therein."

But Aslaug never knew; and, as years went by, the barrow grew until it overtopped the sand-dunes. So went the time until fifteen winters had passed over her.

---

[1] It was a belief that flames hovered at night over the graves of the mighty dead.

Ragnar Lodbrog

RAGNAR AND ASLAUG

Now it fell one day that Ragnar, in his sailing, came nigh to Spangarhede[1] and, seeing there a sandy fjord with pine trees and a spring of fresh water, he sent men ashore to get the water and to bake bread. Then made they an oven of heated stones, and, having drawn forth the fire and set the dough in the oven to bake, they went into the forest. And as they went there came to them the sound of singing; so, treading warily, they followed the sound and reached at last a sunlit pool, where sat Aslaug. As she sang she combed her long hair, shaken loose from the wadmal cap; and, washed clean from stains, her skin shone white as silver in the sun.

Then came they forward and greeted her. And then, for the first time since Heimar went, Aslaug spoke with men.

" Who are ye? " she asked, in no wise afeared. " Whence come ye, and of what people? "

" We are the men of Ragnar of Sweden, and we go a-viking," they answered.

Then bade she them sit near to her while she questioned them of the far lands they had seen; merry was she withal, so that mighty was the sound of laughter in the forest; and it was not until

[1] A point of land in Norway is still called Krakebeck.

towards eventide that they bethought them of their bread.

So they bade farewell to Aslaug and betook them back to their oven, and behold! the bread was burnt black.

Then looked they upon one another, and one said:

"This comes of woman's wiles. Ever will there be trouble where woman is."

And another laughed and said:

"Herein is no blame to the woman, but to the foolish men who had no sense to remember duty when a woman spoke. Still, naught can we do now but bear back burnt bread to the ship."

So they shouldered the bread and bore it seawards to where Ragnar sat on the deck, looking at the sunset.

"Where is the bread?" quoth the forecastle man, "and why bear ye stones upon your shoulders?"

And shamefast they cast down the loaves and said:

"These be bread."

Then did the forecastle man rate them sorely, so that Ragnar came up to see what was to do.

"No man could think on bread with that maiden by," the men repeated; and so much did they tell of her wondrous fairness and wit, that Ragnar said:

"At morn shall ye go and fetch me this maid; and that I may know whether she hath such wit as ye say, bid her come unto me not alone nor yet in company; not clad nor yet unclothed; not fasting yet having eaten naught."

And the men went and, finding Aslaug waiting by the pool, gave the message of the king.

But Aslaug laughed and shook her head:

" How know I thy king's mind? " she asked, " mayhap he might take me and carry me overseas, an I would not. Go ye back and say that I trust no man, and go no whither unless he pledge me by the eye of Odin that I come back scatheless as I go."

Then took they this word unto Ragnar, and first he was wroth that a herd-maiden should doubt him; but in the end, since she would not come without, he sent his gold arm-ring as a pledge for his word.

Then, at red dawn, came a strange sight down the fjord to the long-ship.

Krake, since by this name did the men know her, had bathed in the pool until her skin was white as the winter moonlight; no clothes had she, but a red-brown fishing net was wrapped many times around her and over it, to her knees, showered her golden hair. Naught had she eaten, but she had set her white teeth in an onion, so fasted not. No person came with her, save her dog, so she was not alone.

And when Ragnar beheld her he said:

" Surely no woman had ever wit like unto this maiden's, even as none had ever beauty like unto hers."

And the more he talked with her, the more did he marvel, so that ere evening came he bade her sail away with him and be his wife. But she would not.

" Herd-maidens wed not with kings," she said.

Ragnar and Aslaug

" But here is Thora's robe," he answered; then he sang:

> " Take thou, O sweet, this silver-wrought kirtle,
> Borgar-hjort owned it, and she would rejoice,
> Fain would she, living, have called thee sister,
> Fain would she, dying, have known me in peace.
> Faithful was she till the Nornir divided us,
> Faithful wilt thou be until my life's end."

And Aslaug sang back:

> " Ne'er may I take the silver-wrought kirtle,
> Owned long years since by Thora thy queen;
> Never can eagles mate with the ravens,
> Krake my name is, and coal-black I go,
> Ever in wadmal, herding the cattle,
> Hard must I ever live, far from all wealth."

Since he might not prevail on her at that time to go with him, the king bound her by an oath that, when he had been ten months a-viking, if his mind should still be set upon her, he might return and she would wed him.   Loth was the king to let her go, but, being bound by his word, he led her back to the forest and sailed away from Norway.

" In truth," quoth he, " never have I had so bright a day since Thora died."

Ragnar Lodbrog

Now, when the ten months were overpast, came Ragnar, full of thoughts of Krake, to hold her to her word.

And as the long-ship's sails, striped blue and white, fell upon the deck, and the ship brought-to in the fjord, there on the bank stood Aslaug, fair and white and golden, in the sunlight to welcome them. So she sailed away to Ragnar's land, and there he wedded her, and they lived happily, so that he went no more a-viking.

Four sons had they and two daughters. Of the sons, Ivar, the first born, was strangely made, in that there was no bone in his legs, but only gristle; so that he must ever be borne to war upon a litter of spears. Yet was he wise above all other folks, save only his mother. Next came Björn, and he was a baresark; then Hvirtserk, and young Ragnvald,[1] so called after Ragnar's other son, who had been killed long years before.

Wiser each day grew Aslaug in runes and magic, and it was in her mind that some day might Ragnar, though now aged, wish again to go a-viking, since this had ever been his life's work. Therefore called she her daughters and said:

" Let us make for Ragnar a shirt, wherein I will weave magic and runes so that, old though he be, no steel nor venom shall ever hurt him."

[1] Some versions call him Sigurd.

Ragnar's Sons

And she took silk from the South lands, the fibres of herbs that she alone knew, and some of her hair and that of her daughters, and she taught them magic songs, so, as they wove, they sang, and the spells were worked back and forth through the shirt until they filled it throughout. Then did Aslaug lay it by until it should be needed.

Then did the daughters make a banner also, and sang runes over it; and upon it was worked a raven with great wings that should flap when it was carried to battle.

Now there came one day the King of Upsala to visit Ragnar, and in his mind was a certain plan. This was that Ragnar should put away Aslaug, and take to wife Osten, the king's child; for greatly did he desire a bond between himself and Ragnar, who was first and richest of the Northern kings. And it chanced that they sat together in the court-yard, beneath Aslaug's window, so that she heard all that passed, and how that the king taunted her as a peasant's daughter and no fit mate for the great Ragnar.

But Ragnar put him off with fair words, since he might not flout a guest, and said him neither yea nor nay.

Then was the heart of Aslaug bitter within her, and there came upon her the spirit of Brynhild, her mother, and she spoke no more to the King of Upsala, nor went into the high-hall until he had gone home.

And Ragnar asked:

" Fair wife, what is amiss? "

" Canst thou call her ' fair wife,' who is no jarl's daughter, but a low-born peasant from the forest? "

Then the king knew that she had heard somewhat and told her all; and also how he had put the king's word aside unanswered.

And Aslaug, being blithe that he had hidden nothing from her, told him her story.

"No Krake am I," said she, "but the child of Sigurd, Fafnir's Bane, and Brynhild. The last of the Volsung race am I also, since Swanhild is dead."

And she told all the story of Heimar, as it is here set forth, save that she knew not, even by her runes, how Heimar died, and she ended:

"For myself hast thou loved me, and for thy love ought I to have told thee this before. But I treasured the thought that thy love was given to Krake, and not to the last daughter of the Volsungs. Forgive!"

And Ragnar kissed her joyfully; and word of Aslaug's birth was spread abroad in the land, so that no more did the King of Upsala put forth his daughter, since none could vie with the Volsung race.

Now, during these years had the sons of Ragnar gone a-viking, and many were the deeds they did, and worthy of their sire.

It had befallen, during the life-time of Thora, that Ragnar had made prisoner the King of Northumbria, and forced him to pay scatt[1] yearly unto him. But when this king died, Ella, his son, being young and foolhardy, refused the scatt to Ragnar's jarl; and bade him tell the King of Sweden to come and gather it at point of sword.

[1] Tribute.

Ragnar's Sons

" Bring forth thy magic coat, good wife," cried Ragnar; " once more must I go fight."

And clad in the silken shirt, with the raven-banner floating overhead, did Ragnar Lodbrog sail westward to return no more.

For there were witch-wives in Northumbria, and they, hearing of his cunning, raised a great storm, so that the long-ships were broken in pieces on the rocks, and the king and his men came to shore with naught but their arms.

Then came up Ella with his soldiers against them, and bade them stand and fight; so did they, until all were killed, save the old king who stood alone unhurt amid the ring of dead.

Then said Ella:

" Who art thou that thou comest with warships against me ? "

But Ragnar answered not.

" Nameless canst thou not be," said Ella. " Tell me thy name lest ill befall thee."

Still Ragnar spoke not.

" The worse shall it be for thee if thou wilt not speak," said Ella; but uneasy was he, for he knew that if this were Ragnar, great would be the vengeance of Ragnar's sons, should he be killed.

Yet, since days passed and still Ragnar made no sign, Ella gave word that he should be cast into the orme-gaard, which was a pit full of venomous snakes. " The serpents," said he, " will make him speak. Then, if he be Ragnar, draw him quickly forth."

But the snakes shrank back from the king's

magic shirt, and would in no wise touch him, and, seeing this, the watchers took from him his shirt so that he was bitten on all sides.

Then did Ragnar sing that Death Song that has made his name famous throughout the ages. He told of his battles—fifty-one; of his dead sons; of the kings he had conquered; and he ended:

> " We fought with swords!
> O, that Aslaug's sons
> Knew of their parent's death!
> Valiant is the heart they got
> From their mother, daughter of Sigurd.
>
> We fought with swords!
> Now fades the world away.
> Now comes the call of the gods;
> To one who welcomed it
> Death is a glorious time.
>
> We fought with swords!
> With joy I make an end.
> Fast ride the Valkyrja
> Bearing me to Odin,
> To the halls of Valhalla.
>
> With joy I make an end,
> Soon shall I drink with gods.
> Past are the hours of my life,
> Laughing do I die."

So died Ragnar Lodbrog, greatest of vikings, first even before Rolf the Ganger; and mighty was the vengeance for his death.

Now, when Ella heard of the Death Song he feared greatly, and he sent messengers to the sons of Ragnar, charging them straightly that they should mark well what each son said and did when the news was told, and bring the word to him with all speed.

The Vengeance of Ivar

THE VENGEANCE OF IVAR

Now it chanced that the four brethren were together in the castle hall when there came a warder, running, to say that the messengers of Ella were without.

Then Björn, who stood fixing a spearhead to its handle, laughed:

" It is the scatt," cried he, " Ella hath heard of the setting forth of our father and is afeared."

" Not that, but something worse," said Ivar gravely, from his couch.

Hvirtserk and Sigurd, or Ragnvald his brother, were playing draughts.

And when the messengers were brought in, Ivar spoke them fair and asked their news; and the men told it bravely and without fear, and by no word did the brethren stay them. But when they told of the Death Song and the vengeance that the sons should take, Ivar became by turns pale and red, and his lips were blue; Björn gripped so hard his spear-shaft that his finger-marks were in the wood as long as it lasted; Hvirtserk pressed the draught-piece he held so tightly that blood spirted from his finger-ends; and Sigurd, who had taken up a knife to carve a stick, cut his finger to the bone and knew it not. And for a long space did no one speak when the tale was told.

Then, with an evil laugh, said Hvirtserk:

Ragnar Lodbrog

" Brethren, best were it to kill these men that Ella may know our will."

" Nay," said Ivar, " these men have borne themselves well in a perilous errand, and scatheless shall they return, the more so that I bid them carry this word:

" King of Northumbria, the death words of Ragnar sink deep into the heart of Ragnar's sons, but they bide their time."

So went the men forth with all honour, and carried the message unto Ella. And the king was troubled when he heard it, and said:

" From all Ragnar's sons fear I naught, save only from the still Ivar."

And he made haste to prepare his fighting men, and banded himself with the Kings of England, that he might withstand the onslaught of Lodbrog's sons.

The brethren also made ready ships to go over against Ella, although against the will of Ivar.

" Let be for awhile," said he, " until we know the strength of Ella," and he sent forth messengers to bring word of the doings in Northumbria.

But Aslaug, their mother, brooding fiercely over the death of Ragnar, gave them no peace. Ever sang she the Death Song in their ears as she span, and never passed a day that she did not taunt them, in that their father's death was unavenged.

Then saw Ivar that his brethren were set to go forth, therefore did he give but one warning more:

" Ye know naught of the power of the king," he said, " stay your hand until we gather more news of him, for, I hear, that with him are the Kings of Mercia and Wessex, and, if this be so, we shall but suffer defeat for naught. Prepare well, ere ye go."

" Doth this mean that thou goest not with us? " asked Björn hotly.

" Nay, brother, Ragnar's sons have ever one cause. My men and treasure are yours, and with you I will sail, but I take no part in the fight since I know that ye will fail."

" That is fair speech, brother," said Björn, " with thee there will all go well."

" All will go ill, I tell thee, Björn; yet, if I be there, I may yet remedy it somewhat."

So they set forth in many ships, but as they landed, there came up Ella with a mighty host, and routed them, so that they must go back to their ships, yet not so sorely that they need feel aught but shame at being beaten by the English.

Now Ivar went not back but stayed on land, and sent two jarls unto Ella with this word:

" Were-gild must we have for our father's death, and since I cannot go a-viking with my brethren, I will that thou give me land on this thy shore, that I may dwell in peace."

" That were a fool's trick," quoth Ella, " to set Ragnar's son in the midst of my kingdom."

" Nay," said Ivar, " carry my word to Ella, that never will I bear arms against him, and with the word of Ivar Ragnarsson must he be content."

" And what will Ivar Ragnarsson and his brethren take for were-gild? " asked Ella; " since with Ivar's word, in sooth, I am content."

Ragnar Lodbrog

" Even as much land as he can enclose within an oxhide," the jarls replied.

Then Ella laughed, well-pleased, for he thought:

" This Ivar, of whom we hear great things, is a fool, and can in nowise harm me."

And he granted the were-gild.

But when Björn and his brethren heard of Ivar's doings they were sore angered, and went to Ivar to reproach him. But Ivar would say naught in his own defence.

" Leave me to my own way," he said, " is not our father as much to me as to you? Take from my store at home, each your share of the were-gild and send to me the rest, and "—here he looked narrowly upon each brother in turn—" *when ye get the crossed twig, banded with red, come ye, and your following, with all speed.*"

Then the brethren bade him farewell, knowing some crafty plan must be in Ivar's mind; and sailed away until he should have need of them.

But Ivar bid his men take an oxhide, the largest they could find, and steep it well until it grew lithe; then did he set them to cut it into fine strips, and sew these strips end to end, so that their length was very great. When this was done he sent unto Ella, saying:

" Ivar Ragnarsson desires that thou shouldst be present at the measuring of his land."

So the king came, and greatly was he pleased with the well-favoured face and courteous speech of Ivar, for he had a silvern tongue. And they went forth until they came to a great plain with a fair hill set in the midst; and there Ivar bade them set down his litter.

" This hill would I fain have, O king! "

" Have it and welcome," laughed the king; " as much as thy oxhide will enclose."

Then the thralls brought forth the hide, and the king grew grave when he knew how Ivar had bested him. Yet he made no sign when the whole fair hill, and much land therewith, was compassed round about, for he said to himself:

" Even this is poor pay for the death of such a man as Ragnar Lodbrog."

So Ivar built him a stronghold on the hilltop, which was afterwards called Lincoln, and by reason of his wisdom and justice, there came many to dwell under its shelter, even his countrymen from the Norselands; and his fame as a man of peace went abroad throughout all England. Yet knew no man that Ivar, by his craft, had set the other kings, who were banded with Ella, at odds with him, so that in time the King of Northumbria could depend on none to take his side, save only his own people.

Then came a day when Ivar called the trustiest of his men to him and said:

" Take this, hand it to my brother, Björn, and say unto him, 'Ivar greets thee and his brethren.' "

" Naught but this? " said the man, in wonder, for Ivar had given him but a small stick marked with red.

" Naught but that," he answered, and the man went forth.

Ragnar Lodbrog

Then was great stir in the seaboard of Sweden and Norway, for the brethren had spent these years in making ready, since they trusted Ivar and they never forgot. And word of it came to Ella and troubled him greatly, so that he sent to Ivar, saying:

" An oath didst thou swear unto me, never to bear arms against me. What is this that I hear of thy brethren and their fighting men? If they come up against me, since thou hast thy part in my kingdom, thou must fight on my side."

" Against thee have I sworn not to fight, but against the children of Ragnar and Aslaug, my brethren, neither will I lift an hand."

Then the sons of Ragnar sailed ahead of their ships and came in secret to Ivar. And he told them of the trouble he had made between the kings; of the number of Ella's men; and of where they could best meet him for the furthering of their own ends.

So when their men were come ashore there was a sore fight that lasted throughout the day, and always, when the need was greatest, was Ivar found, borne in his litter, among the Norsemen; yet he held no weapon in his hand, naught but a small white staff.

And at eventide the men of Northumbria were all dead, or fled away, and only Ella remained, a prisoner in the hands of Ragnar's sons.

Then, remembering their father's cruel end, they took him forth next day and slew him, nor did Ivar say a word to stay their hands.

And this should not have been, and is a shame

The Vengeance of Ivar

unto Ivar's name, since he had asked and had been granted were-gild for his father's life.

Yet, in that his deeds were good, and his rule merciful and just, soon did folks forget the death of Ella, and Ivar lived to a great old age within his lands; nor did he ever return to his own country. But his brethren went a-viking, gathering thereby wealth, that Aslaug and their wives warded for them in Sweden, until all perished; but still were left their children to keep alive the name of Ragnar Lodbrog through all days to come.

Printed in Great Britain
by Amazon